Astronauts for Hire
The Emergence of a Commercial Astronaut Corps

Other Springer-Praxis books of related interest by Erik Seedhouse

Tourists in Space: A Practical Guide
2008
ISBN: 978-0-387-74643-2

Lunar Outpost: The Challenges of Establishing a Human Settlement on the Moon
2008
ISBN: 978-0-387-09746-6

Martian Outpost: The Challenges of Establishing a Human Settlement on Mars
2009
ISBN: 978-0-387-98190-1

The New Space Race: China vs. the United States
2009
ISBN: 978-1-4419-0879-7

Prepare for Launch: The Astronaut Training Process
2010
ISBN: 978-1-4419-1349-4

Ocean Outpost: The Future of Humans Living Underwater
2010
ISBN: 978-1-4419-6356-7

Trailblazing Medicine: Sustaining Explorers During Interplanetary Missions
2011
ISBN: 978-1-4419-7828-8

Interplanetary Outpost: The Human and Technological Challenges of Exploring the Outer Planets
2012
ISBN: 978-1-4419-9747-0

Erik Seedhouse

Astronauts for Hire

The Emergence of a Commercial Astronaut Corps

Published in association with
Praxis Publishing
Chichester, UK

Dr Erik Seedhouse, M.Med.Sc., Ph.D., FBIS
Milton
Ontario
Canada

SPRINGER–PRAXIS BOOKS IN SPACE EXPLORATION

ISBN 978-1-4614-0519-1   e-ISBN 978-1-4614-0520-7
DOI 10.1007/978-1-4614-0520-7
Springer New York Heidelberg Dordrecht London

Library of Congress Control Number: 2011936859

© Springer Science+Business Media New York 2012
This work is subject to copyright. All rights are reserved by the Publisher, whether the whole or part of the material is concerned, specifically the rights of translation, reprinting, reuse of illustrations, recitation, broadcasting, reproduction on microfilms or in any other physical way, and transmission or information storage and retrieval, electronic adaptation, computer software, or by similar or dissimilar methodology now known or hereafter developed. Exempted from this legal reservation are brief excerpts in connection with reviews or scholarly analysis or material supplied specifically for the purpose of being entered and executed on a computer system, for exclusive use by the purchaser of the work. Duplication of this publication or parts thereof is permitted only under the provisions of the Copyright Law of the Publisher's location, in its current version, and permission for use must always be obtained from Springer. Permissions for use may be obtained through RightsLink at the Copyright Clearance Center. Violations are liable to prosecution under the respective Copyright Law.
The use of general descriptive names, registered names, trademarks, service marks, etc. in this publication does not imply, even in the absence of a specific statement, that such names are exempt from the relevant protective laws and regulations and therefore free for general use.
While the advice and information in this book are believed to be true and accurate at the date of publication, neither the authors nor the editors nor the publisher can accept any legal responsibility for any errors or omissions that may be made. The publisher makes no warranty, express or implied, with respect to the material contained herein.

Cover design: Jim Wilkie
Project copy editor: Christine Cressy
Typesetting: BookEns, Royston, Herts., UK

Printed on acid-free paper

Springer is part of Springer Science+Business Media (www.springer.com)

Contents

Preface ix
Acknowledgments xi
About the author xiii
List of figures xv
List of tables xix
List of panels xxi
List of abbreviations and acronyms xxiii

Section I The New Generation of Commercial Astronauts 1

1   The birth of the world's first commercial astronaut agency 3
    End of an era 3
    The dawn of a new era 7
    The legacy of SpaceShipOne 9
    Astronauts for Hire is born 12
    Having the right stuff at the wrong time 26

2   Who becomes a commercial astronaut? 29
    Commercial candidates 31
    Prior spaceflight experience is an advantage 33
        Virgin Galactic's requirements for pilot-astronauts 33
    Meet the scientists 35

3   Medical qualification and training 43
    SpaceShipTwo flight profile 44
    Suborbital medical risks 47
        Acceleration 48
        Microgravity effects 52
        Cardiovascular effects 53
        Neurovestibular effects 54
        X-15 neurovestibular experience 54
        Space motion sickness 57

Emergency egress capability 58
Environmental medical issues 59
Radiation 61
Noise 62
Vibration 63
Astronauts for Hire suborbital medical standards 63
Training for commercial suborbital spaceflight 64
NASTAR 65
Astronauts for Hire 68
Orbital flights 71
Orbital medical risks 72
Acceleration 73
Barometric pressure 73
Microgravity 74
Ionizing radiation 76
Noise and vibration 76
Temperature and humidity 76
Behavioral 77
Medical assessment of orbital crewmembers 77
Post-flight medical 80
Orbital training 80
Atlas Aerospace 80
Astronauts for Hire orbital training program 81
References 81

Section II The Industry 83

4 Meet the industry 85
Suborbital operators 92
Virgin Galactic 93
XCOR Aerospace 94
SpaceDev 96
Blue Origin 97
Orbital operators 98
Bigelow Aerospace 99
SpaceX 101
Excalibur Almaz 104
Spaceports 106
Organizations 107
Regulations 108

5 The new breed of space vehicles 111
The new Area 51 111
The X-15 legacy 113
Scaled composites 117

	XCOR	120
	Dream Chaser	123
	Blue Origin	126
	Bigelow Aerospace	127
	SpaceX	130
	Dragon	132
	Boeing CST-100	133

Section III The Missions 139

6 Suborbital missions 141
    Flying a suborbital payload 149
    Launch day 151
    Portfolio of game-changing missions 154
    The suborbital payload agent 157
    Payloads 158
    Getting suborbital research ready for lift-off 161

7 Orbital missions 163
    Orbital infrastructure 164
    Extraterrestrial employment 170
        Space traffic controller 171
        Supervisor astronauts 173
        Orbital entertainment agent 175
        Spacecraft servicing mechanic 176
        Space doctors 177
        Exploration broker 178

8 One small step for entrepreneurship 179
    Mining the Moon 179
        Infrastructure 180
    Mining helium-3 184
    Mining oxygen 187
    Mining ice 188
    Science on the Moon 189
        Astrophysics 189
        Heliophysics 190
        Earth observation 191
        Geology 191
        Materials science 192
    Scientist astronaut duties 192
        Bio-Suit 193
    Commercial Moon operations 195

Appendix I ... 197
Appendix II ... 237
Appendix III ... 239
Index ... 241

Preface

"We're really on the cusp of an exciting new capability for our country and for our economy."
    Lori Garver, NASA's deputy administrator, explaining why NASA is seeking $75 million for NASA's Commercial Reusable Suborbital Research program

As the main engine ignites, the crew feels a deep rumble far below them and a sudden sensation of motion as the launch vehicle lifts off, trailing a 150-m-long fountain of sun-bright exhaust in an inferno of smoke, searing light, and earth-shaking noise. The three crewmembers feel the thunder of the launch, the numbing noise, and the incredible acceleration, as they are pushed forcefully back into their seats. The gut-wrenching journey to orbit – an event planned for many months and anticipated by the crew for several years – takes less than nine minutes. Once in orbit, the thrill of the ascent is replaced by a moment of fulfillment as the spacefarers get their first glimpse of Earth from space – a moment worth a lifetime of anticipation and the hundreds of hours spent training. But this is no ordinary spaceflight. Seated either side of the pilot are two commercial astronauts – astronauts for hire – employed by a research company to conduct experiments in low-Earth orbit.

Until recently, spaceflight had been the providence of a select corps of professional astronauts whose missions, in common with all remarkable exploits, were experienced vicariously by the rest of the world via television reports and internet feeds. These spacefarers risked their lives in the name of science, exploration, and adventure, thanks to government-funded manned spaceflight programs.

All that is about to change.

Section I describes how Astronauts for Hire (A4H) was created in 2010 by Veronica Ann Zabala-Aliberto, Ryan Kobrick, Amnon Govrin, Brian Shiro, and Joe Palaia. Here, the reader is introduced to A4H's vision for opening the space frontier to commercial astronauts and describes the tantalizing science opportunities offered when suborbital and orbital trips become routine. Section I goes on to describe the training and qualification necessary to become a member of the future astronaut corps. The process of acquiring the necessary qualifications to become a government-sponsored astronaut can take 20 years or more. To submit a competitive application, it is generally accepted a candidate must possess a myriad of qualifications, ranging from the requisite Ph.D. and a pilot's license to skydiving credentials and scuba-diving experience. For the budding commercial astronaut, the standards are still challenging, but not as demanding as government selection. Section I concludes by introducing the reader to A4H's suborbital and orbital qualification process, such as the demands of high-altitude indoctrination and the punishing ride in NASTAR's centrifuge.

Section II describes the vehicles that will fly the new crop of commercial astronauts. Anticipation is on the rise for the new fleets of commercial suborbital and orbital spaceships that will serve the scientific and educational market. These reusable rocket-propelled vehicles are expected to offer quick, routine, and affordable access to the edge of space, along with the capability to carry research and educational crew members such as A4H'ers. Yet to be demonstrated is the hoped-for flight rates of suborbital vehicles. Quick turnaround of these craft is central to realizing the profit-making potential of repeated sojourns by commercial astronauts to suborbital and orbital heights. As Section II outlines, vehicle builders still face rigorous shake-out schedules, flight-safety hurdles, as well as extensive trial runs of their respective craft before suborbital space jaunts become commonplace. Section II examines some of these "cash and carry" suborbital craft under development by such companies as Blue Origin, Masten Space Systems, Virgin Galactic, and XCOR Aerospace, and describes the hurdles the space industry must overcome before the hiring of commercial astronauts can develop into a profitable economic entity. It also provides positive suggestions for how the commercial spaceflight industry can plan and prepare for the challenges of marketing and financing the hiring of astronauts. Section II continues by examining the role of commercial operators as enablers of the future of astronauts for hire. It concludes with a vision of a partnership with governments and the private sector and how this collaboration will eventually integrate the free market's innovation of commercial space activities.

Section III describes the various missions this new corps of astronauts will fly and the customers who will employ them. It begins with an assessment of suborbital flights, which may be used to carry out a variety of high-altitude science studies, including access to three to four minutes of microgravity for experimentation in disciplines such as astronomy, life sciences, and microgravity physics. Section III continues by examining the types of missions that will accelerate human expansion outward, beginning with orbital science missions to commercial trips to low-Earth orbit and continuing with Exploration Class missions through cislunar space, the establishment of interplanetary spaceports, lunar bases, and outposts on the surface of Mars. Along the way, it describes the tasks commercial astronauts will perform, ranging from mining asteroids to harvesting helium.

Acknowledgments

In writing this book, the author has been fortunate to have had five reviewers who made such positive comments concerning the content of this publication. He is also grateful to Maury Solomon at Springer and to Clive Horwood and his team at Praxis for guiding this book through the publication process. The author also gratefully acknowledges all those who gave permission to use many of the images in this book, especially Astronauts for Hire members Brian Shiro, Jason Reimuller, Chris Altman, Luis Zea, Veronica Aliberto-Zabala, and Joe Palaia.

The author also expresses his deep appreciation to Christine Cressy, whose attention to detail and patience greatly facilitated the publication of this book, to Jim Wilkie for creating the cover of this book, and to Stewart Harrison, who sourced several of the references that appear in this book.

Once again, no acknowledgment would be complete without special mention of our rambunctious cats, Jasper MiniMach and Lava, who provided endless welcome (and occasionally unwelcome!) distraction and entertainment.

*This book is dedicated to my wife, Doina,
without whom I would never have had the chance of pursuing
my dream of flying into space*

About the author

Erik Seedhouse is a Norwegian-Canadian suborbital astronaut whose life-long ambition to work in space is one step closer to being realized thanks to the organization that provided the inspiration for this book. After completing his first degree in Sports Science at Northumbria University, the author joined the legendary 2nd Battalion the Parachute Regiment, the world's most elite airborne regiment. During his time in the "Para's", Erik spent six months in Belize, where he was trained in the art of jungle warfare and conducted several border patrols along the Belize–Guatemala border. Later, he spent several months learning the intricacies of desert warfare on the Akamas Range in Cyprus. He made more than 30 jumps from a Hercules C130 aircraft, performed more than 200 helicopter abseils, and fired more light anti-tank weapons than he cares to remember!

Upon returning to the comparatively mundane world of academia, the author embarked upon a Master's degree in Medical Science at Sheffield University. He supported his studies by winning prize money in 100 km ultradistance running races. Shortly after placing third in the World 100 km Championships in 1992 and setting the North American 100 km record, the author turned to ultradistance triathlon, winning the World Endurance Triathlon Championships in 1995 and 1996. For good measure, he also won the inaugural World Double Ironman Championships in 1995 and the infamous Decatriathlon, the world's longest triathlon – an event requiring competitors to swim 38 km, cycle 1,800 km, and run 422 km. Non-stop!

Returning to academia once again in 1996, Erik pursued his Ph.D. at the German Space Agency's Institute for Space Medicine. While conducting his Ph.D. studies, he still found time to win Ultraman Hawai'i and the European Ultraman Championships as well as completing the Race Across America (RAAM) bike race. Due to his success as the world's leading ultradistance triathlete, Erik was featured in dozens of magazines and television interviews. In 1997, *GQ* magazine nominated him as the "Fittest Man in the World".

In 1999, Erik decided it was time to get a real job. He retired from being a professional triathlete and started his post-doctoral studies at Vancouver's Simon Fraser University's School of Kinesiology. In 2005, the author worked as an astronaut training consultant for Bigelow Aerospace in Las Vegas and wrote

Tourists in Space, a training manual for spaceflight participants. He is a Fellow of the British Interplanetary Society and a member of the Space Medical Association. Recently, he was one of the final 30 candidates of the Canadian Space Agency's Astronaut Recruitment Campaign. Erik works as a manned spaceflight consultant, professional speaker, triathlon coach, and author. He is the Training Director for Astronauts for Hire (*www.astronauts4hire.org*) and completed his suborbital astronaut training in May 2011. He is eligible for spaceflight assignments and plans to travel into space as an A4H astronaut on board one or more (hopefully several!) of the commercial spacecraft written about in this book.

In addition to being a suborbital astronaut, triathlete, skydiver, pilot, and author, Erik is an avid mountaineer and is currently pursuing his goal of climbing the Seven Summits. *Astronauts for Hire* is his ninth book. When not writing, he spends as much time as possible in Kona on the Big Island of Hawai'i and at his real home in Sandefjord, Norway. Erik is owned by three rambunctious cats – Jasper, Mini-Mach, and Lava – none of whom has expressed any desire to travel into space but who nevertheless provided invaluable assistance in writing this book (!).

Figures

1.1	Space Shuttle mockup inside the Space Vehicle Mockup Facility	4
1.2	Soyuz spacecraft	5
1.3	Retired NASA astronaut Garrett Reisman	6
1.4	Full-scale mockup of Bigelow Aerospace's Space Station Alpha	8
1.5	Artist's rendering of astronauts working on a future Moon base	8
1.6	Astronauts for Hire (A4H) logo	9
1.7	SpaceShipTwo	10
1.8	Interior of SpaceShipTwo	11
1.9	Brian Binnie	12
1.10	Brian Shiro	13
1.11	Flashline Mars Arctic Research Station (FMARS)	15
1.12	Amnon Govrin	15
1.13	Veronica Ann Zabala-Aliberto and Hugh Gregory	17
1.14	Kristine Ferrone and Joseph Palaia operate the Maverick Unmanned Aerial Vehicle	18
1.15	Todd Romberger	20
1.16	Jason Reimuller at the controls of the Space Shuttle mockup	22
1.17	Jason Reimuller preparing to be disoriented	23
1.18	Jason Reimuller preparing for an emergency egress	24
1.19	A4H members undergoing sea survival training at Survival Systems	25
1.20	A4H members before their high-altitude indoctrination training	25
1.21	The author on the summit of Aconcagua, December 2006	27
2.1	Mike Melvill	29
2.2	US Federal Aviation Administration commercial astronaut wings	30
2.3	Astronaut Pin	31
2.4	XCOR's Lynx I rocket plane	36
2.5	Dr Alan Stern	37
2.6	Lockheed F-104 Starfighter	39
2.7	Christopher Altman	40
3.1	Close-up of Alan Shepard inside the Freedom 7 spacecraft	44

Figures

3.2	Technical description of SpaceShipTwo	45
3.3	SpaceShipTwo's flight profile	46
3.4	SpaceShipTwo in a captive flight underneath WhiteKnightTwo	47
3.5	Different types of G	49
3.6	The author being fitted for G-pants	50
3.7	Stoll curve	51
3.8	Neil Armstrong in the X-15 #1 cockpit	53
3.9	X-15 Ship #3 (56-6672) in flight	55
3.10	Michael J. Adams	56
3.11	A4H members conducting emergency egress training	58
3.12	The interior of SpaceShipTwo	59
3.13	The author during a high-altitude indoctrination run	60
3.14	Solar flare activity	62
3.15	Brienna Henwood introduces the NASTAR Center	65
3.16	NASTAR's AFTS-400 centrifuge	66
3.17	Mindy Howard in the NAPSTAR centrifuge	67
3.18	Jose Miguel Hurtado	68
3.19	Laura Stiles	69
3.20	A4H members performing emergency egress training	70
3.21	The author preparing to perform unusual attitude training	70
3.22	Brian Shiro, Christopher Altman, and Jason Reimuller	71
3.23	Robert T. Bigelow	72
3.24	Richard Garriott	79
4.1	SpaceX's Falcon-9	86
4.2	Ariane	87
4.3	SpaceX's Dragon capsule	89
4.4	Boeing's CST-100	90
4.5	Bigelow Aerospace's space station	91
4.6	Artist's rendering of Spaceport America	93
4.7	Flight profile of XCOR's Lynx spacecraft	95
4.8	Sierra Nevada's Dream Chaser	96
4.9	Blue Origin's New Shepard spacecraft	98
4.10	Bigelow Aerospace's BA-330 module	100
4.11	Infographic of SpaceX's Dragon capsule	102–3
4.12	Excalibur Almaz capsule	106
5.1	Mojave Spaceport	113
5.2	Crew members securing the X-15 rocket-powered aircraft while the B-52 mother-ship does a fly-by overhead	114
5.3	X-15 rocket-powered aircraft aloft under the wing of a B-52	115
5.4	X-15	116
5.5	SpaceShipTwo	118
5.6	SpaceShipTwo interior	118
5.7	SpaceShipTwo feathering	120
5.8	Infographic showing the interior of the Lynx	122
5.9	Lynx fitted with cargo pod	122

5.10	Sierra Nevada's Dream Chaser	124
5.11	NASA's HL-20 lifting body mockup	125
5.12	Blue Origin's orbital spaceflight in-flight rendering	126
5.13	Bigelow Aerospace BA-330 modules	128
5.14	Artist's rendering of Bigelow Aerospace's Moon base	130
5.15	The Dragon capsule in orbit	131
5.16	Cutaway of the Dragon capsule showing seating configuration	133
5.17	Cutaway of Boeing's CST-100 capsule showing seating configuration	135
5.18	Liberty launch vehicle	137
5.19	Mobile Launcher	138
6.1	Space Shuttle *Discovery* launches on its 39th and final mission	142
6.2	Zero-G Corporation's specially modified Boeing 727, G-Force One	143
6.3	Black Brant sounding rocket	145
6.4	Suborbital flight suit designed by Orbital Outfitters	152
6.5	Lynx spacecraft carrying the Atsa Suborbital Observatory	155
6.6	Example of a CubeSat	159
6.7	NanoRack	160
7.1	Earth–Moon L1 staging node	164
7.2	Science-fiction space tug	165
7.3	NeoMiner	166
7.4	Artist's illustration of a space-gate	168
7.5	Rendition of an on-orbit fuel depot concept	168
7.6	Three Sundancer modules linked together in space station configuration	170
7.7	Space junk	172
7.8	Dennis Tito	175
7.9	The interior of an orbital medical facility	177
8.1	Artist's rendering of astronauts engaged in mining activities on the Moon	180
8.2	Bigelow Aerospace's lunar base	181
8.3	Sam Rockwell in a still from the movie *Moon*	185
8.4	Mark III Miner	186
8.5	Bio-Suit	194

Tables

6.1	Suggested timeline for a suborbital flight experiment	149
6.2	Suggested events leading up to suborbital science launch	150
6.3	Launch countdown milestones	153
8.1	Sample crew activity schedule	183

Panels

1.1	Your typical A4H flight member: Jason Reimuller	21
5.1	BA-330	129
5.2	Escape systems	134
6.1	Anatomy of a sounding rocket	144
6.2	The stand test	156

Abbreviations

A4H	Astronauts for Hire
A-LOC	Almost Loss of Consciousness
AGSM	Anti-G Straining Maneuver
AMF	Advanced Medical Facility
AsMA	Aerospace Medical Association
CCDev	Commercial Crew Development
CELSS	Closed Environmental Life Support System
COMSTAC	Commercial Space Transportation Advisory Committee
COPD	Chronic Obstructive Pulmonary Disease
COTS	Commercial Orbital Transportation Services
CRS	Commercial Resupply Services
CRUsR	Commercial Reusable Suborbital Research program
CSA	Canadian Space Agency
CSF	Commercial Spaceflight Federation
CSLA	Commercial Space Launch Act
CST	Commercial Space Transportation
CVP	Central Venous Pressure
DOME	Device for Orientation and Motion Environment
EA	Excalibur Almaz
EASA	European Aviation Safety Administration
ECLSS	Environmental Control Life Support System
EML	Earth–Moon Lagrange
ENT	Ear, Nose, Throat
EPU	Environmental Physiology Unit
ESA	European Space Agency
ETC	Environmental Tectonics Corporation
EVA	Extra Vehicular Activity
FAA	Federal Aviation Administration
FAI	Fédération Aéronautique Internationale
FMARS	Flashline Mars Arctic Research Station
G-LOC	Gravity Induced Loss of Consciousness

GCR	Galactic Cosmic Radiation
GOR	Gradual Onset Rate
GSFC	Goddard Spaceflight Center
GST	General Space Training
GTO	Geosynchronous Transfer Orbit
ICRP	International Commission on Radiological Protection
ILOB	Icarus Lunar Observatory Base
ISPCS	International Symposium for Personal and Commercial Spaceflight
ISRU	In-Situ Resource Utilization
ISS	International Space Station
ISU	International Space University
JPL	Jet Propulsion Laboratory
JSC	Johnson Space Center
KSC	Kennedy Space Center
LAS	Launch Abort System
LEO	Low-Earth Orbit
LLOX	Lunar Liquid Oxygen
LPSA	Launch Purchases Services Act
LRO	Lunar Reconnaissance Orbiter
LSS	Lunar Support Scientist
MDRS	Mars Desert Research Station
MIT	Massachusetts Institute for Technology
NASTAR	National Aerospace Training and Research Center
NCRP	National Council on Radiation Protection
NEA	Near Earth Asteroid
NOAA	National Oceanographic and Atmospheric Agency
NPS	Nuclear Power System
NSRC	Next Generation Suborbital Researchers Conference
NTPS	National Test Pilot School
OCST	Office of Commercial Space Transportation
OSC	Orbital Sciences Corporation
OTV	Orbital Transfer Vehicle
PDR	Preliminary Design Review
PFST	Pre-Flight Space Training
PI	Principal Investigator
PSI	Planetary Science Institute
RAF	Royal Air Force
RBS	Reusable Booster System
REM	Research and Education Mission
RLV	Reusable Launch Vehicle
ROR	Rapid Onset Rate
RpK	Rocketplane Kistler
RRV	Reusable Re-entry Vehicle
RSC	Rocket Space Corporation
RTG	Radioisotope Thermoelectric Generator

SAR	Synthetic Aperture Radar
SAS	Special Air Service
SCR	Solar Cosmic Radiation
SEC	Shackleton Energy Company
SFU	Simon Fraser University
SMS	Space Motion Sickness
SNC	Sierra Nevada Corporation
SQM	Strange Quark Matter
SS1	SpaceShipOne
SS2	SpaceShipTwo
SST	Suborbital Scientist Training
SSTP	Suborbital Scientist Training Program
STC	Spaceport Traffic Control
SVMF	Space Vehicle Mockup Facility
SwRI	Southwest Research Institute
TGF	Terrestrial Gamma-ray Flash
TIA	Transient Ischemic Attack
TPS	Thermal Protection System
TRL	Technology Readiness Level
TSC	The Spaceship Company
UAE	United Arab Emirates
UCF	University of Central Florida
ULA	United Launch Alliance
UNISCA	United Nations International Student Conference of Amsterdam
USA	United Space Alliance
USAF	United States Air Force
VSE	Vision for Space Exploration
VTOL	Vertical Take-Off and Landing
WFPC	Wide Field and Planetary Camera
WK1	WhiteKnightOne
WK2	WhiteKnightTwo

Section I

The New Generation of Commercial Astronauts

1

The Birth of the World's First Commercial Astronaut Agency

Johnson Space Center (JSC), Houston. Inside a cavernous hangar, a full-scale replica of the Space Shuttle fuselage (Figure 1.1) rests like a beached whale on a floor almost two football fields long. It shares the space with a mockup of the International Space Station (ISS), a Soyuz capsule, smoke machines, and a myriad other tools of the astronaut-training trade. The hangar, aka the Space Vehicle Mockup Facility (SVMF), is where astronauts spend a lot of their time training and operating equipment they will use in orbit. As they enter the SVMF, they pass through a dividing wall that displays mug shots of assigned Shuttle crews, with the Shuttle team closest to launch placed in the prime position on the wall. Tim Reynolds, who serves as the Operations Control Center Supervisor, knows it's a big deal for astronauts to get their photo on the wall; he's managed the SVMF through 62 Shuttle flights and 27 ISS crews. In his long career with NASA, he's seen many an astronaut come and go for simulation training, but, with the Shuttle retired, the queue for the prime spot on "Tim's Wall" won't be as long as it used to be.

THE END OF AN ERA

You see, the retirement of the Space Shuttle spells the beginning of the end of the US astronaut corps; 50 years after its birth, the astronaut program – an iconic American venture – is undergoing a transformation. The program probably won't vanish. Probably. But, now that the astronauts' main ride into space has been retired and funding for true manned space exploration shrinks to almost nothing, the astronaut corps will become smaller, its role redefined, and more of the space duties will likely be turned over to commercial spaceflight companies. It's a transformation that will morph a corps once dominated by test pilots to one increasingly made up of scientists and specialists who will fly into space, conducting experiments and flying payloads. This new band of spacefarers will still be pioneers, although they won't be elevated to hero status like the astronauts in the 1960s and 1970s and it's also unlikely they will grace the covers of magazines.

4 The Birth of the World's First Commercial Astronaut Agency

Figure 1.1 A Space Shuttle mockup inside the Space Vehicle Mockup Facility. Courtesy: NASA

For those who follow the trajectory of manned spaceflight, the prospect of NASA's laying off astronauts shouldn't have come as a surprise. After all, NASA had been preparing for the end of the Shuttle program and a downsized astronaut corps since January 2004, nearly a year after the *Columbia* disaster. At the time, President George W. Bush unveiled his vision for space exploration, which called for terminating the Shuttle program in 2010, ending US involvement in the ISS in 2015, and developing Ares I,[1] which could deliver a crew of four to low-Earth orbit (LEO) by 2014. The vision, dubbed Constellation, even called for US boots on the Moon by 2020. To bridge the gap between the Shuttle's retirement and the first launch of Ares I, NASA bought seats on Russia's Soyuz (Figure 1.2) craft (at $61 million dollars each!) to deliver US crews to the ISS. In the meantime, the US continues to work on a replacement for the Shuttle – an objective that will most likely be achieved by commercial enterprise thanks to President Obama's controversial effort to establish a space program that is financially sustainable.

---

[1] Ares I was the crew launch vehicle that was being developed by NASA to launch Orion, the spacecraft intended for human spaceflight missions after the Space Shuttle was retired. Ares I was to complement the larger, unmanned Ares V, which was intended as the cargo launch vehicle of President Bush's space exploration vision, which was canceled in October 2010 by the passage of the 2010 NASA authorization bill.

Figure 1.2 Soyuz spacecraft. Courtesy: NASA

It was on April 15th, 2010, that President Obama made the case for the most radical change of direction in NASA's history, wading into a debate laden with emotion, big bucks, and ambitions that united a nation in a race to the Moon more than four decades ago. Under Obama's fiscal 2011 budget request, NASA has now exited the business of blasting astronauts into orbit at least in the near term. Instead, Obama is gambling that nascent aerospace firms will develop the Shuttle's replacement quicker and more cheaply than NASA. If he's right, the president would provide a huge boost to the space entrepreneurs described in this book, potentially unleashing the greatest new engine for innovation since the internet. Using rockets to blast humans into LEO is 50-year-old technology, the Obama administration contends. The administration argues that, rather than tying up NASA resources, such flights should be outsourced to NASA partners such as Boeing and upstarts like SpaceX, the rocket company founded by internet magnate Elon Musk. Writing off the $9 billion already spent to return to the Moon and eventually land humans on Mars, Obama proposed to boost NASA's $19 billion budget by $6 billion over the next five years, with most of the increase used to seed commercial development. Of course, critics claim that, without an exploration program to keep NASA's goals in focus, the agency's research will devolve into a science fair whose funding eventually will be gutted by Congress without fear of political repercussions, but that's a subject for another book.

While NASA still plans on building a rocket that can make the trip to the ISS and perhaps send astronauts to an asteroid and eventually Mars, the numbers of astronauts required for these missions will be far smaller than in the Shuttle era. In short, there will be far fewer mug shots on Tim's wall. From a peak of more than 100 astronauts in 2000, the roster as this book is being written stands at just 61 and is dwindling (the corps is expected to fall to 51 flight-eligible astronauts by 2016); 15 years ago, NASA selected 10 pilot-astronauts and 25 mission specialists, but in 2009, it selected just nine for its two-year astronaut basic training program. Further reductions may be on the cards. Like an employer phasing out an old product line and trying to figure out what it will do next, NASA Administrator and former astronaut Charles Bolden Jr tasked the National Research Council's Committee on Human Spaceflight Crew Operations to recommend changes to the role and size of the astronaut corps; the board, backed by Chief Astronaut Peggy Whitson, believes NASA will need 55–60 astronauts in this post-Shuttle era – with such a small roster, Whitson expects future astronaut selection campaigns to select small classes in the four to six range rather than the double-figure classes that were common in the 1980s and 1990s (in September 2011, NASA announced an astronaut selection campaign to be held in 2012).

Not surprisingly, as opportunities to leave the planet shrink and budgets tighten, astronauts have been weighing the pros and cons of whether to leave the agency. Some, like Garrett Reisman (Figure 1.3), have already left and joined private rocketeers such as SpaceX, which is under contract with NASA to provide unmanned (and eventually manned) resupply flights to the ISS. Others are keeping their spacesuits on, hoping for an ISS slot.

Figure 1.3 Retired NASA astronaut Garrett Reisman spacewalks outside the International Space Station during the outpost's Expedition 16 mission. Courtesy: NASA

The private sector also offers the opportunity for people to become astronauts who didn't come up through NASA's ranks. Take David Mackay, for example. Mackay is Virgin Galactic's chief pilot. He grew up in a remote village in northern Scotland, where the military frequently ran training flights. Captivated with becoming an astronaut after watching the Apollo moon landings, he followed the usual path to try to become an astronaut: working as a military test pilot. It was only when he reached his 30s that it dawned on him that he was never going to make it as an astronaut if he stayed in the UK, a country that only recently had a British candidate selected as an astronaut (Timothy Peake, via the European Space Agency's – ESA's –2008 astronaut selection campaign). After leaving the Royal Air Force (RAF), Mackay switched over to life as a civilian pilot, piloting Boeing 747s and Airbus 340s for Virgin Atlantic. He's now setting up Virgin Galactic's astronaut program and will be one of the first pilots of the passenger rocket, SpaceShipTwo (SS2), and its jet-powered mother ship, WhiteKnightTwo (WK2).

Mackay will be one of the very first commercial astronauts and the demand for people like him will grow as private companies expand beyond just ferrying high-end tourists into space. For example, in late February 2011, the Southwest Research Institute (SwRI) signed contracts with Mackay's company, Virgin Galactic, and California-based XCOR Aerospace, another company building craft for suborbital flights; the flights, which may number as many as 17 launches, will focus on scientific research. It was a deal put in place largely due to the efforts of Dr Alan Stern, a former associate NASA administrator, planetary scientist, and an associate vice president with the non-profit institute. Stern firmly believes that entrepreneurial space firms will eventually carry out hundreds of flights per year – a flight rate that will dramatically reduce the cost of doing suborbital experiments, which, in turn, will create jobs for all sorts of commercial astronauts.

THE DAWN OF A NEW ERA

For those ardent fans of the space program, it's not difficult to imagine a future in which commercial inflatable space stations (Figure 1.4) circle Earth, nor is it too much of a stretch to envisage Moon bases (Figure 1.5) dotted on the lunar landscape, populated by a new corps of commercial astronaut. Now, we're not talking about events that will happen 20 or 30 years from now; commercial enterprises such as Bigelow Aerospace intend to be flying sovereign customers on board inflatable habitats by 2015, with a Moon base to follow shortly after. But who will look after those customers? Who will maintain the experiments? And who will pack the latest batch of protein crystals for their return trip to Earth?

Well, no doubt Bigelow will hire and train a core group of astronauts to help his sovereign customers, but what about when other operators start flying *their* customers into orbit? After all, Richard Branson, owner of Virgin Galactic, has already stated his plans of building an orbiting hotel. Perhaps Bigelow and Branson will decide to subcontract astronaut services or perhaps an organization can provide astronaut services at a lower cost? Fortunately, such an organization already exists

Figure 1.4 A full-scale mockup of Bigelow Aerospace's Space Station Alpha inside their facility in North Las Vegas. Courtesy: Bigelow Aerospace

Figure 1.5 Artist's rendering of astronauts working on a future Moon base. Courtesy: NASA

Figure 1.6 Astronauts for Hire (A4H) logo. Courtesy: A4H

and its name is Astronauts for Hire (A4H). Currently, A4H's service (Figure 1.6) is limited to training the next generation of suborbital space pioneers, but training is already being developed to prepare their members for work on orbit and beyond. What follows is how this pioneering organization was created, but first let's look back at the catalyst of the commercial spaceflight revolution.

> "Anything is possible, which is what is exciting. At the right price level, there will be enormous quantities of space ships and space ports around the world."
>
> Richard Branson, commenting after the unveiling of his space vehicle in December 2009

THE LEGACY OF SPACESHIPONE

October 4th, 2004. A momentous event is taking place at Mojave Airport, a sprawling test center in the California desert, just 150 km from Los Angeles. Under the blazing sun, hundreds of rusting aircraft, their engines and undercarriages shrink-wrapped, sit parked in long, lonely rows. But, on this Monday morning, the motley collection of DC10s, 747s, DC9s, and 737s will bear witness to a truly extraordinary event. Here, at this desolate airport, a small, winged spacecraft built with lightweight composites and powered by a rocket motor using laughing gas and rubber will fly to the edge of space and into the history books. Registered with the Federal Aviation Administration (FAA) by the alphanumeric designation N328KF,[2] but known to space enthusiasts as SpaceShipOne (SS1), this privately developed

---

[2] SS1 was unveiled at the Smithsonian Institution's National Air and Space Museum on October 5th, 2005, in the Milestones of Flight Gallery. It's now on display to the public in the main atrium between the Spirit of St Louis and the Bell X-1. The whole project cost less than US$25 million – about the same as NASA spends every day before lunch!

10 The Birth of the World's First Commercial Astronaut Agency

Figure 1.7 SpaceShipTwo. Courtesy: The Spaceship Company

spacecraft (Figure 1.7) will have a galvanizing effect on the nascent commercial spaceflight industry.

The excitement began building the night before, as cars poured into the parking lot and continued to stream in almost until take-off, by which time crowd-control personnel had almost given up. Rows of trucks with satellite dishes and glaring spotlights greet the spectators as they stream into the airport. It's only five in the morning, but a sense of expectancy already wafts through the air together with the smell of coffee and bagels. A huge Ansari X-Prize banner flutters from the control tower as thousands of space enthusiasts from around the world wait for the appearance of N328KF. Buzz Aldrin and other legends of the space program rub shoulders with William Shatner and Mojave's maverick engineering genius, Burt Rutan. Only a few kilometers away, at Edwards Air Force Base, on August 22nd, 1963, test pilot Joe Walker reached the edge of space by flying an Air Force X-15 rocket plane to an altitude of 107,333 m. The X-15 eventually gave birth to the Space Shuttle, a semi-reusable vehicle embroiled in politics that became a symbol that the high frontier was the sole dominion of governments and space agencies – a status quo perpetuated for more than three decades. Until now.

Like all of Rutan's creations, the world's first private spacecraft is an impressive feat of engineering. Marked by simplicity of design, the vehicle doesn't look like it should fly into space. The interior (Figure 1.8) is spare and devoid of the myriad switches, dials, and toggles that crowd the Space Shuttle flight deck. There are a few low-tech levers, pedals, and buttons that suggest the vehicle is designed to fly, but the austere design doesn't scream "space".

The legacy of SpaceShipOne 11

Figure 1.8 Interior of SpaceShipTwo. Courtesy: The Spaceship Company

"WhiteKnight is taxiing" crackles over the public address system. The announcement is followed by the sound of high-pitched jet engines that mark the arrival of a gleaming white carrier aircraft. Slung tightly underneath is SS1. WhiteKnight and SS1 take off from Runway 30 at 0647 local time, followed by two prop-driven chase planes; they will follow SS1 during its one-hour ride to separation altitude, giving spectators plenty of time to grab another bagel and a coffee.

"Three minutes to separation" comes the announcement. Spectators scan the sky, searching for the thin white line that is SS1. At the drop altitude of 14,000 m, SS1 is released and dropped like a bomb above the Mojave Desert. Falling wings level, pilot and soon-to-be astronaut, ex-Navy Test Pilot Brian Binnie, 51, trims SS1's control surfaces for a positive nose-up pitch. Then, he fires the rocket motor, boosting the spacecraft almost vertically. "It looks great," says Binnie as he rockets upwards at Mach 3. Within seconds, SS1 is gone, trailing a stream of white smoke. SS1 accelerates for more than a minute, subjecting Binnie to three times the force of gravity. At 45,000-m altitude, the engines shut down and SS1 continues on its ballistic trajectory to an altitude of 114,421 m. A loud cheer erupts from the spectators, who are following the proceedings on a giant screen, each of them euphoric with the realization that high above them is a spacecraft that may one day carry them into space. With his spacecraft's rear wings feathered to increase drag upon re-entry, Binnie prepares to bring SS1 back to Earth. On the ground, the

Figure 1.9 Brian Binnie. Courtesy: The Spaceship Company

spectators wait, spellbound, straining to hear the double sonic boom that will announce SS1's return. Seconds later, the unmistakable sound announces that SS1 is on her way back from her historic mission. Binnie guides SS1 gently back to Earth, gliding the spacecraft back to a perfect touchdown on the runway. He has just become the 434th person to fly into space (Figure 1.9). Welcoming him enthusiastically upon landing are nearly 30,000 spectators, Microsoft's cofounder, Paul Allen, who helped finance the project, Burt Rutan, SS1's designer, and Peter Diamandis, chairman of the Ansari X-Prize Foundation.

ASTRONAUTS FOR HIRE IS BORN

Three months later in St Louis, Burt Rutan was presented with the Ansari X-Prize. Among the spectators was a young astronaut wannabe, Brian Shiro (Figure 1.10). Inspired by a love for science fiction, Shiro had enrolled in college, intending to be an astrophysicist with the goal of joining NASA's astronaut corps. But, when his university cancelled the astronomy major, he ended up triple-majoring in integrated science, geological sciences, and physics instead. Never losing sight of his dream of one day flying in space, Shiro worked as an intern at NASA's JSC and spent two summers at the agency's Goddard Space Flight Center (GSFC) and Jet Propulsion

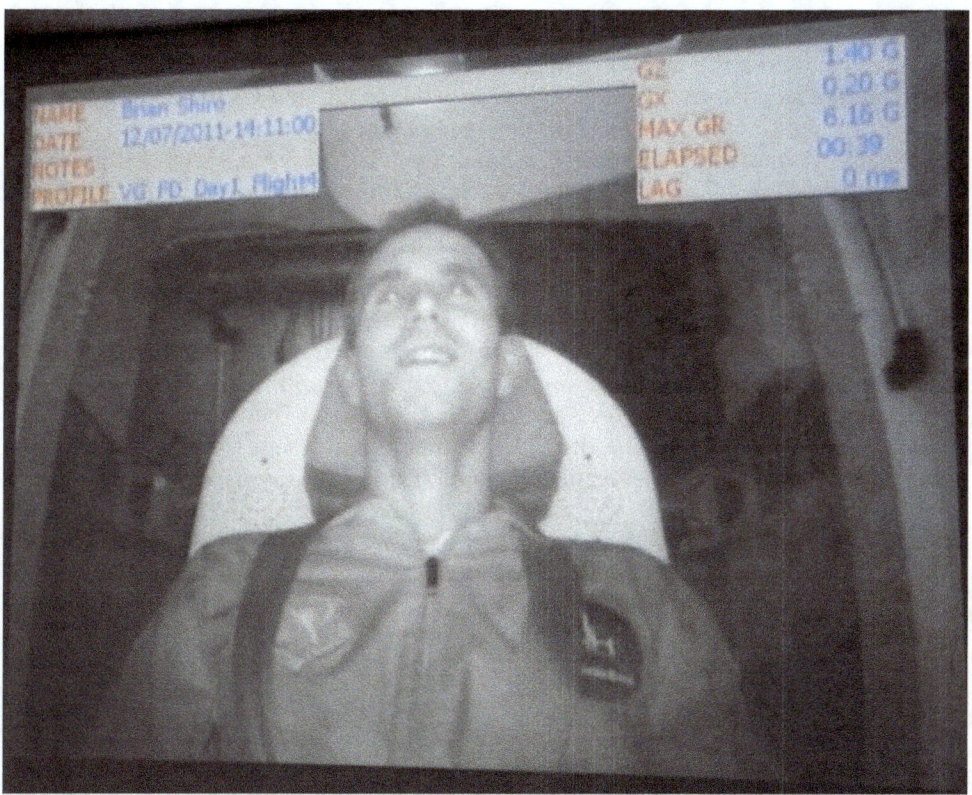

Figure 1.10 Brian Shiro, president of A4H, pulling G inside NASTAR's centrifuge. Courtesy: A4H

Lab (JPL). Upon graduating, he worked as a geophysicist, visiting exotic locations that included Tonga, Fiji, Antarctica, and the Northern Mariana Islands.

Shiro had just returned from yet another field expedition when he witnessed the Ansari X-Prize presentation. With the dream of becoming an astronaut very much on his mind, Shiro arranged a meeting with Peter Diamandis at an Explorer's Club meeting. Diamandis encouraged Shiro to enroll in the International Space University (ISU) as a way of boosting his already impressive résumé. The following year, Shiro found himself in Vancouver for the ISU's Summer Session Program – an experience that opened his eyes to the potential of the commercial spaceflight industry. Although the genesis of a commercial spaceflight corps had yet to take hold, the building blocks were falling into place.

Galvanized by his ISU experience, Shiro finished his Master's degree in Earth and Planetary Science at Washington University with the intention of continuing on to the Ph.D. program. A Ph.D. is almost a pre-requisite for those hoping to be selected as NASA mission specialists, but, unfortunately, politics, personal, and funding hurdles torpedoed his Ph.D. chances, so Shiro took a job with the National Oceanographic and Atmospheric Agency (NOAA). For the next two years, with no astronaut-hiring campaign on the horizon, he worked as a geophysicist, bought a house, and started a family. Then, in 2007, NASA announced they would be hiring a fresh batch of astronauts. In February 2008, while compiling his NASA application, Shiro happened to catch a *Good Morning America* broadcast live from the National AeroSpace Training and Research Center (NASTAR). NASTAR, a high-tech aerospace training facility located in Southampton, Pennsylvania, had just announced its partnership with Richard Branson's Virgin Galactic company and was touting its first suborbital scientist class. Thinking a ride in a centrifuge and some high-altitude experience in a hypobaric chamber might help his chances of astronaut selection, Shiro called Brienna Henwood, who was in charge of NASTAR's training program, and she described their training program. Unfortunately, the cost of NASTAR's training was too much, so Shiro submitted his NASA application without the centrifuge and hypobaric checks. Then, like the other 3,535 applicants, he waited. Thinking other applicants might be interested in his selection experience, Shiro tracked the details on his blog called "Astronaut for Hire", which he had started in late 2007 to chronicle his experiences of NASA's astronaut selection. By January 2009, he had been identified by NASA as a "Highly Qualified" candidate and his blog was a hit among the elite astronaut candidate community. Sadly, Shiro didn't make the final cut. Disappointed but undeterred, he volunteered as a crewmember for a simulated Mars expedition at the Flashline Mars Arctic Research Station (FMARS – Figure 1.11) on Devon Island in the Canadian Arctic. It was a fortuitous decision because it was on Devon Island that Shiro met many who were to become members of an organization set to revolutionize the commercial spaceflight industry.

Following FMARS, Shiro returned home to Oahu, where he had another fortuitous encounter when he met astronaut Dr Yvonne Cagle, who suggested he attend the upcoming inaugural Next Generation Suborbital Researchers Conference (NSRC). "You'll love it," Cagle said. "You'll learn all about the NASA CRUsR[3] program and you'll meet a lot of really cool people doing really cool stuff in the commercial space industry." Shiro didn't need to be persuaded; if Cagle, who was managing the CRUsR program at the time, thought that NSRC was a good bet, it was a good bet. Deciding to attend NSRC, Shiro started reading about the suborbital space community and looked forward to learning more at the conference. At the same time, more than 5,000 km away, in Boulder, Colorado, Amnon Govrin, another astronaut hopeful, was also investigating the potential of the commercial space industry. A 36-year-old software engineer and father of three, Govrin (Figure 1.12) had followed Shiro's "Astronaut for Hire" blog with interest and was thinking of creating his own blog. He decided to give Shiro a call.

---

[3] In 2009, NASA's Innovative Partnerships Program (IPP) established the Commercial Reusable Suborbital Research Program (CRuSR) Office at Ames Research Center (ARC). The program (since renamed "Flight Opportunities") is intended to investigate the potential procurement of reusable suborbital spaceflight services and solicit research investigations that will utilize commercial reusable suborbital spaceflight services.

Figure 1.11 Flashline Mars Arctic Research Station (FMARS). Courtesy: Mars Society

Figure 1.12 Amnon Govrin outside the Mars Society Desert Research Station. Courtesy: Amnon Govrin

Govrin, an Israeli engineer, had served in the Israeli Defense Force for three years, during which he had written and decoded software for X-ray imaging and improved ways to measure the effects of bullets. After his military service, Govrin worked for Mercury Interactive, before being sent to Boulder, Colorado, to spearhead the integration of a newly acquired company called Freshwater Software. Then, in 2004, he started work for Webroot Software, a provider of security software; after helping the company build the enterprise endpoint protection product, he became a software development manager for the main consumer product line of the company. In that function, he developed and released the 2011 security product line, which received the *PC-Magazine* Editors' Choice Award. He now works in Seattle as the Software Development Manager for Amazon Video on Demand and as Chief Technology Officer for A4H.

Govrin and Shiro discussed ways to become more involved in the commercial spaceflight industry. Shiro was preparing to attend NSRC and was happy to help. Then, shortly after talking with Govrin, Shiro got a call from the Mars Society;[4] they had selected him to command a two-week mission at the Mars Desert Research Station (MDRS) in Utah. The MDRS mission conflicted with NSRC, but Shiro couldn't pass up an opportunity; Shiro immediately called Govrin and suggested he visit him while he was in Utah. Govrin accepted. A few weeks later, Shiro, true to his word, drove the pressurized rover from the MDRS site into nearby Hanksville and collected Govrin and his kids. While driving around the MDRS site in the Mars buggy, Govrin mentioned he was planning on attending NSRC. Shiro suggested he introduce himself to Veronica Ann Zabala-Aliberto ("Vaza"), one of Shiro's MDRS colleagues.

Vaza (Figure 1.13) had served on Crew 36 on the first all-female crew at a Mars analogue station in March 2005. In addition to being the Founder and President of the Phoenix Chapter of the National Space Society (NSS), Vaza also serves as a JPL Solar System Ambassador, served on the National Space Society Board of Directors from 2006 to 2008, and worked on the Lunar Reconnaissance Orbiter (LRO) mission that launched in Spring 2009. At NSRC, Vaza met up with Amnon at the session presenting the NASTAR Suborbital Scientist Training Program (SSTP) and immediately started talking about how to pool their networking resources so they could find sponsors to help cover the $6,000 cost of the training.

The inaugural NSRC was a huge success, bringing the research and education communities together with suborbital vehicle providers and government funding agencies to explore the new era of suborbital spaceflight. When the idea of NSRC was first conceived, many people were skeptical such a meeting would generate much interest because they thought the research and education communities just wouldn't

---

[4] The Mars Society is an international space advocacy non-profit organization dedicated to promoting the human exploration and settlement of the planet Mars. Founded by Robert Zubrin and others in 1998, the organization has attracted the support of notable science fiction writers and filmmakers, including Kim Stanley Robinson and James Cameron.

Figure 1.13 Veronica Ann Zabala-Aliberto (left) and Hugh Gregory (right), a spaceflight historian, climb a hill outside the Mars Desert Research Station in June 2005, several miles north-west of Hanksville, Utah. The station based in the Utah desert is sponsored by the Mars Society and is used for scientific research and practice for a manned mission to Mars in the future. There are several groups a year that man the station one to two weeks at a time. Courtesy: George Frey/Getty Images

be interested in suborbital science. NSRC 2010 proved the naysayers wrong, with people voting with their feet and attending by the hundreds. The excitement among the attendees present at the technical sessions and discussion panels was contagious and proved a galvanizing experience, especially for the A4H founders.

Shiro had followed the proceedings via e-mail from Utah. The day after NSRC ended, he received a call from Vaza. She described the need for scientists to "babysit" payloads for researchers and NASTAR's SSTP. She asked Shiro if he was interested in signing up. Shiro replied that he was, but, with a baby at home, an arctic expedition to pay off, and ongoing graduate school expenses, Shiro just didn't have the money. He mulled over the situation for a few days, thinking how he and the others could generate the NASTAR cash. Why not scholarships? He decided to bounce the idea off a fellow FMARS crewmember, Joe Palaia. Like all budding astronauts, Palaia (Figure 1.14) has a fascinating résumé.

A recipient of the Edward Bloustein Distinguished Scholarship and the New Jersey Institute of Chemists Distinguished Award, Palaia was also a graduate of the Massachusetts Institute of Technology's (MIT) Plasma Science and Fusion Center, where he had worked on the Plasmatron project, a device that could turn foodstuffs into fuel for cars. During his time at MIT, Palaia also attended a reactor design class,

18 The Birth of the World's First Commercial Astronaut Agency

Figure 1.14 Crew members Kristine Ferrone and Joseph Palaia operate the Maverick Unmanned Aerial Vehicle on July 24th, 2009. Courtesy: A4H

which investigated how to use nuclear reactors for Mars missions. The class published a report of their work, "Mission to Mars: How to Get There and Back with Nuclear Energy", in 2004 and presented their findings to NASA's Project Prometheus staff. In a related study, Palaia joined other scientists and engineers to engage in an eight-month pre-design study of the first permanent settlement on Mars. In 2005, with his focus still firmly on Mars, Palaia co-founded a space commerce company, the 4Frontiers Corporation; he's served as Vice President of Operations and R&D ever since. The same year, he attended the Space Generation Congress and International Astronautical Congress in Fukuoka, Japan, where he received the Peter Diamandis Leadership Award in recognition of his contributions to the work of the Space Generation. After graduating from MIT with his Master's in Nuclear Science and Engineering in 2006, Palaia became a full-time employee of the 4Frontiers Corporation, continuing his work to make a manned mission to Mars a reality. In July 2009, he served as executive officer and chief engineer for a one-month simulated Mars mission at FMARS.

Shiro and Palaia had been FMARS crewmembers together, so he was familiar with the 4Frontiers Vice President. Shiro pitched Palaia his idea of bringing a group of people to NASTAR funded by scholarships. Palaia, a serial entrepreneur, immediately saw the business potential and suggested they form a non-profit organization to channel donations into a general fund that could then be used to award scholarships to members of the organization. Sitting in his backyard in Ewa Beach, Oahu, the organization began to crystallize in Shiro's mind. Palaia suggested a points system, allowing members to earn points for qualifications such as advanced degrees, training increments, and the amount of time they devoted to the organization; the points would be a way of giving scholarship money to the members who applied for it. Shiro agreed it was an excellent idea. A few days later, he spoke with Henwood and relayed the conversation he had had with Palaia. Henwood was immediately impressed with what Shiro and Palaia were trying to do and said if they could persuade 10 people to take the SSTP course, they would qualify for an academic rate. Shiro started contacting his FMARS and MDRS colleagues immediately. One of the first to respond was fellow A4H co-founder, Ryan Kobrick. Kobrick, a four-time veteran of MDRS (crews 25, 44, 56, and 58), had accumulated impressive academic credentials. He had followed up his Master's of Space Studies degree from the ISU with another Master's degree in Aerospace Engineering, which focused on portable life-support systems. Not content with only two Master's degrees, he gained a Ph.D. for good measure, studying the rather esoteric subject of bioastronautics while at the University of Colorado (he was also an interviewee in the Canadian Space Agency's (CSA's) 2008 Astronaut Recruitment Campaign). In a matter of days, Vaza, Govrin, and Palaia had also signed up. They, in turn, contacted their colleagues and, very soon, the group numbered 11. While the group was assembling, Shiro and Palaia continued to strategize how best to make the organization work. On March 9th, Shiro was checking his inbox when one of Palaia's e-mails caught his eye:

> "Why not put a PR spin on this? Create a website and social media following. Pick seven people for the first 'astronauts for hire'. Issue a press release. Make it reminiscent of the announcement of the original Mercury 7 astronauts. Seek sponsors to cover the costs of training. When the commercial industry is ready to launch, we'll be ready. And we'll inspire a whole new generation to do it.
> Just an idea."

It was more than just an idea; Palaia's suggestion was the genesis for the creation of A4H. Shortly afterwards, the group created a website, set up Google Apps' suite of tools to help collaboration, and, after some debate, chose the name "Astronauts for Hire" – a reflection of Shiro's blog but with an expanded scope (incidentally, Kobrick is credited with suggesting pluralizing "astronaut"). Palaia incorporated the organization in Florida and, on 12th April, 2010, A4H was born as a Florida non-profit corporation (Appendix I). Interest was immediate; in the first month alone, A4H received more than 50 serious expressions of interest from budding astronauts for hire (the author of this book included!).

The A4H team spent the summer of 2010 writing by-laws, establishing a Board of Directors, and electing officer roles. Gradually, the structure of A4H took shape. The team also created two types of members: Flight Members (Panel 1.1) – eligible for training subsidies and scholarships – and Associate Members. As the organization took shape, A4H began to attract media interest. In July 2010, *New Scientist*, the world's leading science and technology news publication, published an interview with Brian Shiro. While the *New Scientist* piece introduced A4H to the scientific community, it was the organization's first job that really put them on the map. In September 2010, hot on the heels of the final launch of Space Shuttle *Endeavour*, a Reuters article entitled "Australian Beer Hopes to Boldly Go into Space" began making waves in the news. The article alluded to the announcement that Vostok Pty Ltd and Saber Astronautics had hired A4H to conduct microgravity flight tests of their new space beer. The honor of conducting the beer test went to Todd Romberger, a veteran of more than 300 parabolas in zero-G aircraft; the fact he was an amateur beer brewer also helped! After some flight schedule changes, A4H successfully carried out the first space beer test on February 26th, 2011, completing their first research contract as "astronauts for hire". The flight (Figure 1.15) spurred a windfall of stories (not surprising really, since beer and space are such popular topics!), with *Fox News*, *Discovery News*, *MSNBC*, *Slashdot*, *FOOD Magazine*, *CBS News*, *Fox News*, PRI's *The World*, *Sydney Morning Herald*, *Toronto Sun*, *The Independent*, *Daily Telegraph*, *Daily Mail*, *The Guardian*, *New Scientist*, *Forbes*, *MSNBC*, and *Discovery News* featuring the flight. Principal Investigator of the experiment Dr Jason Held had this to say about A4H:

"We hired Astronauts4Hire to support our first microgravity flight experiment because of the demonstrated experience of their team and their ability to accomplish the job. A4H delivered everything with consummate professionalism and diligence, with the attention to detail expected from medical and aerospace field research. Thank you A4H!"

Figure 1.15 A4H member, Todd Romberger, tests the world's first space beer on board the Zero-G aircraft, February 2011. Courtesy: A4H

> **Panel 1.1. Your typical A4H flight member: Jason Reimuller**
>
> Jason Reimuller (Figure 1.16) is A4H's Chief Operating Officer. Like all A4Hers, he's aspired to be an astronaut as long as he can remember. A recent Ph.D. graduate in Aerospace Engineering Sciences, Jason is an experienced system engineer and project manager and recently founded Integrated Spaceflight Services, based on prior experience he gained while conducting abort-scenario and emergency-egress analyses for NASA's Constellation Program. In addition to his experience working with NASA astronauts and mission planners, Jason accumulated a wealth of engineering experience via his work with satellite propulsion systems at Space Systems Loral (his exposure to spacecraft operations came first as a space command officer in the US Air Force, serving later as a satellite flight director for Space Systems Loral) and at the Sondrestrom Incoherent Scatter Radar Facility in Greenland.
>
> Not content with just one Master's degree (there are no under-achievers in A4H!), Jason holds M.S. degrees in Physics *and* Aviation Systems. He's also a recipient of the NASA Graduate Student Research Fellowship, a multi-engine, instrument-rated pilot, and conducts research in glaciology to better understand the dynamic changes of the Greenlandic ice sheet. In his spare time, he scuba-dives, practices his Spanish, and studies Russian.

Unfortunately, despite all the media interest, donations were not forthcoming and A4H had to decide how to improve fundraising. To open doors for tax-exempt fundraising meant gaining federal 501(c) (3) status, which, in turn, meant lots of paperwork – more than 100 pages of it. The application, which required the organization to project detailed cost and revenue projections for the next three years, was submitted in January 2011 and approved six months later. In parallel with the 501(c) (3) application, the A4H team decided to expand the organization to help with the workload and paperwork! A Flight Member recruitment campaign announced in December 2010 received more than 50 applications, from which the selection committee (which included two former NASA astronauts) chose nine new members.

As this book is being written, A4H has a complement of more than 60 members representing 17 countries across North America, Europe, Asia, and Latin America. Many of the A4H members were finalists or highly qualified candidates in recent NASA, the CSA, and the ESA astronaut selection campaigns; all of them are accomplished scientists and engineers representing a wide variety of fields, including aerospace engineering, aerospace medicine, aviation, biology, chemistry, electrical engineering, genomics, geology, geophysics, industrial engineering, mechanical engineering, materials, medicine, neurobiology, oceanography, nuclear engineering, physics, planetary science, space architecture, space science, and sports medicine. As the world's only commercial astronaut training cooperative and flight crew service,

22 The Birth of the World's First Commercial Astronaut Agency

Figure 1.16 Jason Reimuller at the controls of the Space Shuttle mockup. Courtesy: A4H

A4H provides opportunities for astronaut candidates to develop and refine the skills necessary to become commercial scientist-astronauts and helps those individuals find flight opportunities as payload or mission specialists. In consultation with former astronauts and astronaut trainers, the organization has developed a comprehensive astronaut medical and training certification program to qualify its members for spaceflight missions. In July 2011, the inaugural A4H class completed the NASTAR SSTP, which included spatial orientation tasks (Figure 1.17), emergency egress drills (Figure 1.18), sea survival training (Figure 1.19), hypobaric indoctrination (Figure 1.20), and centrifuge training; soon, this group will be fully fledged trained astronauts eligible for flight assignments. As a Research and Education Affiliate of the Commercial Spaceflight Federation and a sponsor of the 2011 NSRC, an organization that started out as a blog has established itself as an integral player contributing to the growth of the commercial spaceflight industry. Going forward, A4H will position itself as the industry's main facilitator of astronaut crew training and development.

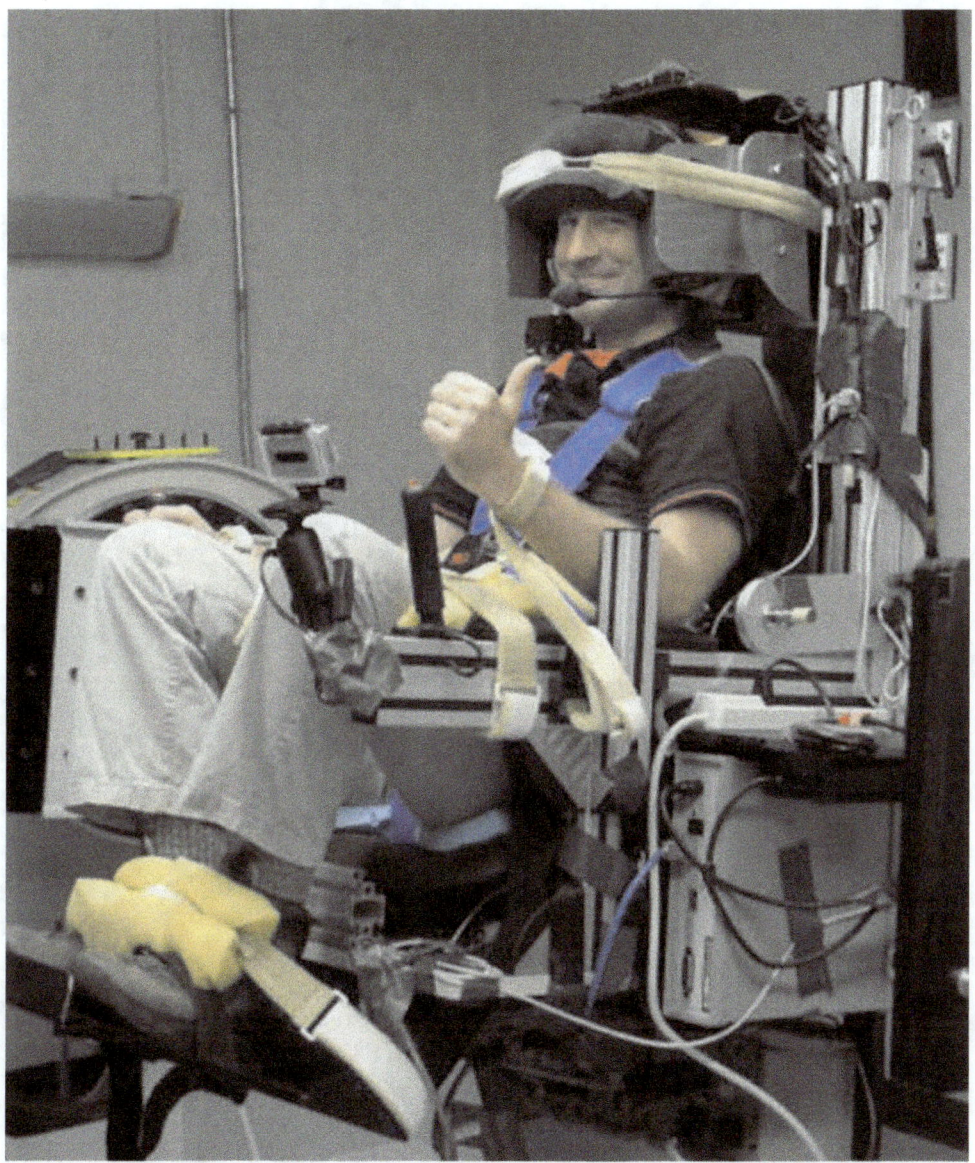

Figure 1.17 Jason Reimuller preparing to be disoriented. Courtesy: A4H

Figure 1.18 Jason Reimuller preparing for an emergency egress. Courtesy: A4H

Astronauts for Hire is born 25

Figure 1.19 A4H members (from left to right) Jules, Chris Altman, Brian Shiro, and Jason Reimuller undergoing sea survival training at Survival Systems, July 2011. Courtesy: A4H

Figure 1.20 A4H members (from left to right) Bran Shiro, Luis Miguel Hurtado, Kristine Ferrone, Alli Taylor, Chris Altman, Jason Reimuller, and Jules Ung before their high-altitude indoctrination training. Courtesy: A4H

HAVING THE RIGHT STUFF AT THE WRONG TIME

Like Brian, Amnon, Ryan, Veronica, and Joe, I've always wanted to be an astronaut. Every kid growing up during the space race wanted to be an astronaut. I have a vivid memory of sitting on the lounge floor, watching the first lunar landing on a black-and-white television set; we didn't have color back in my day! After resolving to become an astronaut, I learned more about what was required to be selected and focused increasingly on achieving my dream.

In 1987, knowing many astronauts had military experience, I joined the 2nd Battalion of the Parachute Regiment. During my time in the Paras, I was trained by Special Air Service (SAS) soldiers in the art of jungle and desert warfare, made night jumps from Chinook helicopters, and jumped out of a Hercules C-130 dozens of times. I also endured annual 80-km marches with a 20-kg pack on my back and acquired advanced survival techniques in the jungles of Belize and the desert of northern Cyprus.

After leaving the Paras, I stepped back into the comparatively mundane (just about any job would be mundane after you've spent time in the Paras – trust me!) life of academia to start my Master's in Medical Science. Shortly after completing my Master's degree, I applied to the Canadian Astronaut Program. Since I hadn't accumulated the myriad qualifications required for a competitive application, my objective was to indicate to the CSA my interest in becoming an astronaut, with the intention of being able to submit a more competitive application when the next selection was advertised. Little did I know that the next selection would be 16 years in the future!

Between the 1992 and 2008 selections, I checked a few more boxes in the astronaut list; I participated as a subject in the ESA's 22nd Parabolic Flight Campaign, I won World Triathlon Championships, and gained my Ph.D. at the German Space Agency thanks to a $50,000 ESA grant. I also checked off accelerated free-fall, a pilot's license, mountain-climbing (Figure 1.21), a post-doctoral degree at Simon Fraser University's (SFU) Environmental Physiology Unit (EPU), and worked as an Astronaut Training Consultant for Bigelow Aerospace. So, when the call for applications finally came in May 2008, I figured I had as good a chance as anyone, even though the odds were high; the CSA received more than 5,000 applications for just two positions. But, after a year of selection tests, including sea survival, fighting fires, robotics assessment, and a myriad other trials and evaluations, I didn't make the final cut (I made it to the final 30). For those applicants who had dedicated 20 years or more to the goal of becoming an astronaut the blow was devastating to put it mildly. For many of those deselected at the final stage who had waited 16 years since the CSA's 1992 campaign, it was the end of the road. The lucky few who worked in the US indicated they would apply for a Green Card with the intent of applying to NASA at the next campaign. Others returned to their jobs in academia, flying jets, or performing research. A small number found themselves recalibrating their life, trying to find another challenge; one of the most highly qualified candidates confided to me that he was considering climbing K2. Some reflected on the high price they had paid in their personal lives (many had

Having the right stuff at the wrong time 27

Figure 1.21 The author on the summit of Aconcagua, December 2006. Courtesy: Author

burned through more than one marriage) by following a dream demanding sacrifices few can fathom.

I was setting my sights on an Everest expedition as a way of getting over the disappointment of not being selected when I came across Brian's blog and sent him an e-mail. He suggested I talk to Vaza, since I was going to be in her home town of Phoenix for the 2010 Aerospace Medical Conference (AsMA). A day after returning from AsMA, I submitted my application to join A4H. Two months later, I was voted in as Training Director. I don't have my ticket into space. Yet. But, by being a member of an organization that is set to reinvigorate interest in manned spaceflight, I *will* fly, just as all the A4H members will, because the commercial spaceflight industry is taking flight and this book explains how.

2

Who Becomes a Commercial Astronaut?

"One of the most important things about exploring space is that it is something that sets us apart in this time that makes history for our nation and for our society."

Dr Alan Stern

Until very recently, the only astronauts were those trained by either the military or civilian space agencies. That changed on June 21st, 2004, when the US Federal Aviation Administration (FAA) issued the first pair of commercial astronaut wings to Mike Melvill (Figure 2.1) following the final test flight of SpaceShipOne (SS1).

Figure 2.1 Mike Melvill receives his astronaut wings after the flight of SpaceShipOne. He is now known as the first civilian to fly an aircraft into space and the first private pilot to earn astronaut wings. The veteran test pilot, with over 7,000 hours of flying time in more than 130 types of aircraft, made headlines in June 2004 when he flew SpaceShipOne at Mach 3 beyond Earth's atmosphere, soaring into space. After a 90° roll and a malfunctioning motor, Melvill stabilized SpaceShipOne during his first flight. He had another white-knuckle ascent on his second trip, fighting to regain control of the craft as it rolled more than two dozen times. Courtesy: Scaled Composites, LLC

30 Who Becomes a Commercial Astronaut?

Figure 2.2 The US Federal Aviation Administration has granted commercial astronaut wings to civilian pilots who have performed a successful spaceflight. Currently, only Mike Melvill and Brian Binnie have these wings, but that is going to change very soon. Courtesy: FAA

On that historic day, Melvill flew to an altitude of 328,491 ft, or about 100 km. Three-and-a-half months later, ex-Navy aviator Brian Binnie became the second commercial astronaut when he flew SS1 to an even greater altitude. Binnie, a test pilot and graduate of the US Naval Test Pilot School (NTPS), flew many different aircraft while in the military, but it wasn't until he began flying commercially for Scaled Composites as a test pilot that he had the opportunity to become an astronaut. Melvill's and Binnie's story is one that will become more and more common.

While the identities of the first batch of commercial astronauts are not yet known, the likelihood is they'll be flying for Virgin Galactic. Envisioned by Sir Richard Branson, Virgin Galactic will be flying the first generation of suborbital spacecraft fresh off the Scaled Composites assembly line. Hot on the heels of Virgin Galactic may be another group of commercial astronauts flying the Dragon, currently being designed and manufactured by a company called SpaceX. And then there's Astronauts for Hire (A4H), the established market leader in commercial astronaut recruitment. As more and more commercial ventures start operations, the space industry will witness the ranks of commercial astronauts swell to meet the demand. And that demand will rise quickly. Take a good look at the commercial astronaut wings in Figure 2.2; the next time you see them, they may just be pinned on the chest of your pilot.

The FAA wings were the idea of Michelle S. Murray, an aerospace engineer who worked in the agency's Office of Commercial Space Transportation (CST), tasked with regulating businesses aiming to travel to sub-orbit and beyond. The pins resemble those worn by NASA astronauts, but the eligibility requirements for wearing the pins are not the same for NASA astronauts and their commercial counterparts. You see, if you happen to be a NASA astronaut, you're awarded two pins: a silver one is awarded to candidates who have successfully completed astronaut training and a gold pin is awarded to those who have actually flown in space. Incidentally, astronaut candidates are given silver pins but are required to purchase the gold pin (Figure 2.3) at a cost of approximately $400. Which brings us to the question: will passengers also qualify for astronaut wings?

According to Hank Price, a spokesman for the FAA, the answer is "No";

Figure 2.3 An Astronaut Pin is issued to all NASA astronauts. It is a lapel pin, worn on civilian clothing. The pin is issued in two grades – silver and gold – with the silver pin awarded to candidates who have successfully completed astronaut training and the gold pin to astronauts who have flown in space. Incidentally, Alan Bean (Apollo 12) took his silver pin to the Moon and left it on the lunar surface, explaining that, since he had worn the silver pin for six years and he would be wearing a gold pin after the mission, he wouldn't be needing the silver one anymore. Courtesy: NASA

commercial wings will only be awarded to the pilot and the crew. That seems a bit harsh if you've paid $200,000 for a ticket, but those are the rules. And, if you think the pilot may take their hands off the controls so the passenger can say they've acted as a pilot, think again; a pilot who did that would violate their license and that would be the end of their commercial astronaut career. Now, you may be wondering why all the astronauts who flew on the Shuttle received astronaut wings, even though only two astronauts piloted the spacecraft. Well, according to NASA rules, the criteria for being a professional astronaut are more for participation than sitting in the pilot's seat. If you think about it, that's only fair. After all, professional astronauts have spent decades preparing themselves for spaceflight and have spent years training for a mission, while a passenger who's paid a couple of hundred thousand dollars for the experience has spent just three or four days training. Whether these rules will be changed to take into account astronauts for hire, who will be paid to work as payload specialists, remains to be seen; after all, this class of spacefarer will be gainfully employed as an astronaut and will receive more extensive training than passengers who pay for a suborbital joyride; the more likely solution will be that organizations such as A4H will award their own astronaut wings to their members.

COMMERCIAL CANDIDATES

"I saw what was going on in the commercial space area and I really believe strongly this is the future of human spaceflight and I had a strong desire to get involved with that. That's really what led me to make this decision, which was not an easy one to make."
NASA astronaut Garrett Reisman explaining why he quit the agency to join up-and-coming private spaceflight company SpaceX

Now that we've established who may qualify as a commercial astronaut, we'll move on to the subject of who these commercial astronauts may be. We can be pretty sure these pilots will be astronauts for hire, but what backgrounds will they have? Will the new crop of commercial astronauts be drawn from NASA retirees or will they be recruited from the ranks of those applicants who missed the cut during astronaut selection? The answer, as we'll see in the following section, is a bit of both.

As a NASA astronaut, Garrett Reisman flew on the Space Shuttle twice, spent three months living on the International Space Station (ISS), and floated in space on spacewalks. In 2011, at the age of 43, he decided to hang up his spacesuit to take a job with Space Exploration Technologies (SpaceX), a commercial space enterprise that plans to launch astronauts (government and commercial) on its Dragon capsule and Falcon-9 rocket. For Reisman, walking away from NASA was a difficult decision but, if he had stayed with the agency, his chances of a third flight would have been almost non-existent – there were just too many astronauts vying for too few flights. At SpaceX, Reisman works with Ken Bowersox, another former NASA astronaut. They work in the area of astronaut safety and mission assurance, which is all about making sure the Dragon and the Falcon-9 are as safe as possible before the new corps of commercial astronauts fly on board.

The departure of Reisman and other NASA astronauts inevitably led to the question: "How many astronauts does NASA need?" That's the question that was set before a special committee[1] (see Chapter 1) tasked by the White House to conduct a 10-month study on the appropriate role and size of the 64-member astronaut corps after the final Shuttle mission. Let's face it; it was only a matter of time before the bad economy and the increasing political pressure to cut government spending would lead to this question being asked. After all, now that the Shuttle era is mothballed, there aren't many flight opportunities for those NASA astronauts that remain (more than half of the 64 current astronauts are without a scheduled mission); with more than 30[2] ISS crew slots yet to be assigned for missions through 2020, some observers are questioning whether less expensive astronauts for hire might be able to fill those roles. Financially, such a question makes sense but, while astronauts for hire may eventually become the norm, for the near term at least, it will be ex-NASA and fighter pilot types who will be the first commercial astronauts to be recruited. That's because space operators need experienced pilots to develop and test their spacecraft, and who better to do that job than a retired agency astronaut? That's certainly the thinking at Virgin Galactic, who advertised for their first batch of pilot-astronauts in 2010.

---

[1] The Committee on Human Spaceflight Crew Operations is funded by NASA under the auspices of The National Academies. It comprises an ad hoc 15-member group of individuals, six of whom are former NASA astronauts, including Bonnie Dunbar and Richard Covey.
[2] If NASA decides to reduce tours of duty at the space station from six months to three, that would mean a need for more astronauts, but that decision has yet to be made.

PRIOR SPACEFLIGHT EXPERIENCE IS AN ADVANTAGE

As you can see from reading Virgin Galactic's astronaut requirements, the standards for applying are almost as stringent as NASA's and it's a situation that's unlikely to change any time soon. That's because – at least in the first few years of the commercial space industry – commercial pilot-astronauts will be charged with the testing and development of the company's vehicles – a task requiring myriad skill-sets.

Virgin Galactic's requirements for pilot-astronauts

Pilot-astronauts must:

- be a US citizen;
- have a current FAA commercial (or equivalent) pilot license and FAA medical;
- have a degree-level qualification in a relevant technical field;
- be a graduate of a recognized test pilot school, with at least 2.5 years' flight test experience;
- have a diverse flying background with a minimum of 3,000 hr flying to include considerable experience of: large multi-engine aircraft and high-performance fast jet aircraft and low lift-to-drag ratio glide experience in complex aircraft;
- have operational experience of an aerospace aviation project or business;
- have excellent, current knowledge on a diverse range of aerospace matters;
- have the ability to communicate aviation knowledge and safety-related information simply, succinctly, and clearly;
- have previous responsibility for authorizing and implementing policies and procedures to ensure a safe and efficient operation;
- be a proven team player;
- Preference will be given to those with experience of: spaceflight, commercial flight operations, and flight instruction.

For example, commercial pilot-astronauts will be responsible not only for regular flight duties on board the spacecraft, but also for flight-planning activities, Mission Control co-ordination, and ground-to-air support, record-keeping, and training obligations. Most likely, they will also be responsible for training new company pilot-astronauts, which will involve supervising ground and flight training and implementing improvement programs for pilot-astronaut training. They may also be involved in training activities with revenue-paying astronauts prior to flight to ensure familiarity and competence with communications systems, seat and harness mechanisms, and general flight procedures. The initial corps of Virgin Galactic pilot-astronauts will also be tasked with establishing the recruitment standard for future pilot-astronauts and assisting in the selection and training of additional commercial pilot-astronauts. Then, once commercial operations have commenced, the pilot-astronaut will move from the role of testing the vehicle to flying the missions.

34 Who Becomes a Commercial Astronaut?

Not surprisingly, the standards for those working in the orbital realm are even more exacting than suborbital requirements. For example, Bigelow Aerospace, whose astronauts will be looking after the company's sovereign customers once it has its space station on orbit, had the following requirements in its advertisement:

Bigelow Aerospace Position Opening: Astronaut
Description
Bigelow Aerospace seeks professional astronauts to fill permanent positions. Qualified applicants need to have completed a training program from their government or recognized space agency and have at least some flight experience on a recognized space mission. Specialized training and/or experience (ie: Medical, Payload Specialist, EVA, Pilot, etc.) is not a pre requisite, but is definitely a plus. Possible opportunities will come in two areas:

1) *Ground*
- Working with the Marketing Team to secure government and cooperate with clients.
- Working with the Design and Fabrication Teams to help optimize layout of systems for on-orbit serviceability and ergonomics.
- Working with the Mission Control Team on final checkout of flight vehicles, both pre and post launch.
- Help develop Astronaut training programs for Bigelow Aerospace Professional Astronaut Corps as well as Client Astronaut Corps.
- Work instructing in the Bigelow Astronaut Training Program.

2. *Flight*
- Perform as Professional Astronaut aboard Bigelow Aerospace Station Complex.
- Manage all on-board aspects of employee and customer astronaut personal safety.
- Maintain the Station Complex as required (mainly IVA, but some EVA as well).
- Help clients with payloads or experiments (primarily with regards to integration into station's systems and communications).

As this book was being written, Virgin Galactic announced it had selected its first astronaut pilot, Keith Colmer. Colmer, a former Air Force test pilot, has logged over 5,000 hours of flight time on 90 different types of aircraft. He also worked on the development and operations side at the Air Force, both for aircraft and for the USAF spacecraft program. As a Virgin Galactic astronaut, he will begin training to fly Virgin Galactic's SpaceShipTwo (SS2). For an astronaut, making the transition from flying the Space Shuttle to flying a suborbital mission may be akin to trading your Audi R8 for a Corolla or going from piloting a Formula 1 car to driving in the NASCAR series. For those accustomed to being in charge of more than 6,000,000 pounds of thrust, working in the suborbital realm may be a bittersweet adjustment – at least until companies like Boeing and SpaceX can develop their more powerful orbital spacecraft. But don't feel too bad for the astronauts; after all, they'll still be doing the coolest job there is and at least they'll be flying while their colleagues back at NASA face the uncertainty of competing for a handful of slots on the ISS. Also, the realities of an agency without a vehicle – NASA astronauts will be guests on board Russian Soyuz capsules for the foreseeable future – have meant morale is not what it used to be at NASA. But, while government astronauts wrestle with the new realities of dampened career aspirations and jockey for the coveted flights, the astronauts for hire who have seen the writing on the wall will be where they want to be: on the verge of a golden age of spaceflight.

But is that really how astronauts for hire will evolve? Let's take a look at some other possibilities. One situation is that the astronaut corps will remain essentially the same. In this scenario, NASA will hire commercial spacecraft and rockets to undertake their planned missions and will use their own team of astronauts to carry out the missions. It's possible, but, given the current pace and transformation of the commercial space sector, such a scenario is unlikely. Another possibility is that NASA and the commercial sector will have its own ranks of astronauts. This would be like the government and the private sector having their own ranks of pilots to fly aircraft. In this scenario, NASA would use their astronauts to fly certain exploration missions, while the private sector would use their astronauts for hire to fly more routine – and mundane – missions such as resupply flights and space tourist missions. A third, more extreme possibility is that NASA will get out of the astronaut business altogether and contract flight crews from the private sector in the same manner as they hire private firms to design and build spacecraft. In this scenario, astronauts would be hired to complete missions on an as-needed basis. As I said, it's unlikely that NASA will rely on astronauts for hire, which leaves us with the second scenario as being the most realistic. But, whichever scenario occurs, it's likely that – in the near term at least – there will be competition between ex-NASA astronauts and upcoming commercial astronaut candidates such as those being trained by A4H. However, don't forget that we're in an austere fiscal environment and astronauts are extremely expensive to train.[3] Also, many of the government-trained astronauts will be nearing retirement so, in the long term, the private astronaut candidates stand to inherit the industry. Plus, the high level of training and skills NASA astronauts possess likely means they will cost more to hire than up-and-comers. I'm willing to bet there will still be jobs for today's aspiring astronaut candidates in the first generation of private spacecraft, since they can charge less for their services.

MEET THE SCIENTISTS

In February 2011, the commercial space industry booked its first science expeditions.

---

[3] A NASA inspector general's report criticized the inefficient use of highly trained astronauts. Part of the reason they're so expensive is because they fly so rarely, which means they must be retrained before the next launch. This training involves more than just classroom work and expensive computer simulators. A fleet of T-38 jets is maintained so astronauts can keep up their proficiency at high-performance jet operations and giant water pools are used to train astronauts for spacewalks.

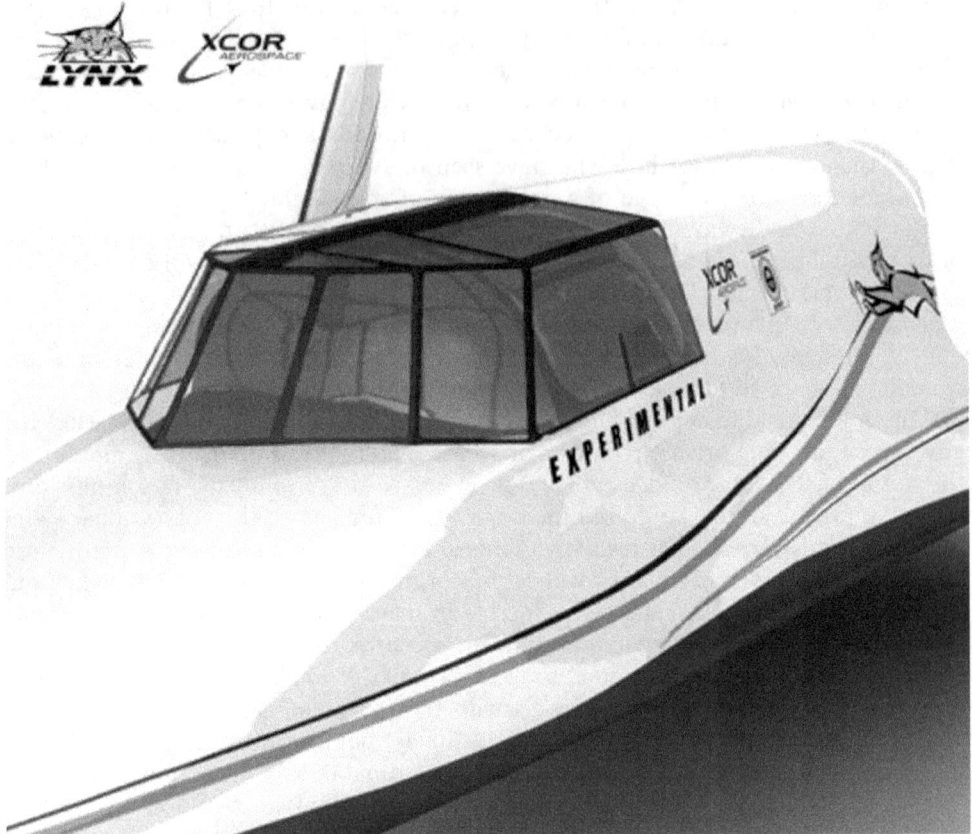

Figure 2.4 XCOR's Lynx I rocket plane. Courtesy: XCOR

Unlike the fanfare that accompanies news of space tourism, the announcement went largely unnoticed in the media, which was a shame because, 10 years from now, those science flights will probably be judged as being one of the most significant events in the commercial space industry. As a result of the $1.6 million deal between the Southwest Research Institute (SwRI) and Virgin Galactic, three researchers booked tickets to fly on Virgin Galactic's SS2, with another six seats on reserve. SwRI also reserved six flights on XCOR's Lynx I rocket plane (Figure 2.4).

The deal was largely thanks to the work of Dr Alan Stern, SwRI's associate vice president for space science and engineering. With the end of the Shuttle era, scientists like Stern were running out of ways to get their experiments in space. Small CubeSat missions were an option, but such automated experiments are very limited in scope, so Stern and his SwRI scientists thought why not buy tickets on suborbital vehicles and fly the experiments themselves?

"We are strong believers in the transformational power of commercial, next-generation suborbital vehicles to advance many kinds of research. We also believe that by putting scientists in space or the upper atmosphere with their experiments, researchers can achieve better results at lower cost and with a higher probability of success than with many old-style automated experiments."

Dr Alan Stern, SwRI's associate vice president for space science and engineering

Stern (Figure 2.5), an astrophysicist and planetary scientist, is a seasoned, hard-charging principal investigator who served as NASA's associate administrator for science until March 2008, when he abruptly resigned from the agency after ruffling

Figure 2.5 Dr Alan Stern, Associate Vice President, Space Science and Engineering Division, Southwest Research Institute, stepping from the cockpit of the F104 Starfighter. Dr Stern is a planetary scientist and an author who has published more than 175 technical papers and 40 popular articles. His research has focused on studies of the Kuiper belt and Oort cloud, comets, satellites of the outer planets, Pluto, and the search for evidence of solar systems around other stars. He has worked on spacecraft rendezvous theory, terrestrial polar mesospheric clouds, galactic astrophysics, and studies of tenuous satellite atmospheres, including the atmosphere of the Moon. He was the principal investigator of the Southwest Ultraviolet Imaging System, which flew on two Space Shuttle missions – STS-85 in 1997 and STS-93 in 1999 – and has been a guest observer on numerous NASA satellite observatories, including the International Ultraviolet Explorer and the Hubble Space Telescope. Stern holds Bachelor's degrees in Physics and Astronomy and Master's degrees in Aerospace Engineering and Planetary Atmospheres from the University of Texas, Austin. In 1989, Stern earned a doctorate in astrophysics and planetary science from the University of Colorado at Boulder. Courtesy: Southwest Research Institute

feathers with fiscal belt-tightening.[4] He now works as an Associate Vice President of SwRI and as the principal investigator of NASA's New Horizons mission to Pluto. In common with all budding astronauts (in 1995, he was selected as a NASA mission specialist finalist and, in 1996, he was a candidate payload specialist but didn't have the opportunity to fly on the Shuttle), he has an impressive résumé, having been a researcher in 24 suborbital, orbital, and planetary space missions, and led the development of eight scientific instruments for planetary and near-space research missions.

Despite a stellar résumé as a planetary scientist, Stern will probably be remembered more as a visionary entrepreneur; not only did he kick-start suborbital science via the Next Generation Suborbital Researchers Conference (NSRC); he also opened the door to space for scientists who may otherwise never have had an opportunity to fly.

Flying with Stern will be two SwRI scientists, Cathy Olkin and Dan Durda. Both Olkin and Durda are planetary scientists; Olkin has an interest in outer solar system worlds, while Durda's research involves smashing granite balls together to study asteroids. With their suborbital flights on the horizon, all three have been busy training in zero-G aircraft, centrifuges, and Starfighter F-104 jet fighters (Figure 2.6).

Their trailblazing flights will be the first of many; the SwRI researchers hope to leverage the initial flights (during which they will conduct biomedical, microgravity, and astronomical imaging experiments) into a wide range of additional flight projects in next-generation suborbital spaceflight stretching out across this decade and the next. After the SwRI flights, others will follow with their experiments and payloads – scientists like Christopher Altman (Figure 2.7).

Suborbital astronaut, scientist, and diplomat: Altman's research focus – quantum coherence and entanglement – is the stuff of science-fiction novels. He began his career at the aptly named Starlab, a multidisciplinary deep future institute, whose projects included everything from time travel and quantum computing to protein folding and nanoelectronics. After peering into the future at Starlab, he was elected to serve as Chairman of the United Nations International Student Conference of Amsterdam (UNISCA) First Committee on Disarmament and International Security. While working as Chairman, he led several hundred upcoming diplomats in an effort to combat international terrorism, ensure global and regional nuclear security, and shape the long-term future of information security. Next, he was assigned to canvas East Asia and create postdoctoral-level national assessments of quantum technology for US policy and funding agency directors at NSA and DARPA under the ATIP Quantum Information Science and Technology Project. For his contributions to the field, he was selected to receive the 2004 RSA Information Security Award for Outstanding Achievement in Government Policy.

---

[4] On taking office, he pledged a reinvestment in research and analysis as one of the main thrusts of his strategy for getting more out of NASA's flat science budget. To that end, he created some anguish among researchers and contractors by turning back requests for more money from projects experiencing cost overruns, forcing mission leaders to trim parts of their projects or find other sources of financing.

Figure 2.6 The Lockheed F-104 Starfighter is a single-engine, high-performance, supersonic interceptor aircraft originally developed for the United States Air Force (USAF) by Lockheed, although NASA flew a small fleet of F-104 types in supersonic flight tests and spaceflight programs until they were retired in 1994. The aircraft shown here belongs to Starfighters Aerospace, which has teamed up with Incredible Adventures to offer civilian suborbital flight training in the Mach 2 aircraft. Pricing starts at US$30,000 for a single flight training program, although, by adding US$23,000, you can have a second flight. Courtesy: Starfighters Aerospace

Returning to the US, Altman completed zero-G and high-altitude physiological training as part of NASA's Reduced Gravity Research Program at Ames Research Center in Silicon Valley and NASA's Johnson Space Center (JSC) in Houston. He was then tasked to represent NASA Ames at the joint US–Japan space conference (JUSTSAP) and the launch conference (PISCES) for an astronaut training facility on the slopes of Mauna Kea volcano on the Big Island of Hawai'i, where next-generation spacesuit prototypes, augmented reality interfaces, and robotics field tests are planned for the next series of manned lunar and Martian missions.

An experienced glider pilot, motorcycling enthusiast, scuba-diver, and rock climber, Altman also speaks Japanese, Dutch, and basic French, is proficient in muay thai, kendo, and judo, and holds a first-degree black belt in kyudo, a form of Japanese archery. With credentials like these, it wasn't surprising that he was selected as an astronaut candidate in A4H's second round of astronaut selection.

So, commercial astronauts could be scientists like Alan Stern and Dan Durda, tasked by their employer to conduct research, or they could be like Christopher Altman, employed by a research organization via A4H. They could also be NASA scientists; while this group may not qualify as commercial astronauts, they are worth mentioning due to the collaborative nature of the commercial spaceflight industry. Let me explain.

Thanks to the regular missions, suborbital flights will provide an opportunity to develop and demonstrate technology and instruments that will later

Figure 2.7 Christopher Altman prepped for his high-altitude indoctrination training at NASTAR. Courtesy: A4H

fly on NASA spacecraft. By conducting in-flight suborbital test and evaluation, scientists would experience cradle-to-grave hands-on mission experience and training. To get the ball rolling, NASA could restore its suborbital program as a foundation of its mission responsibilities, workforce requirements, and instrumentation development needs. It could then increase funding to suborbital programs and assign a program lead to the staff of the associate administrator for the Science

Mission Directorate to coordinate the suborbital program. This lead would be responsible for the development of strategic plans for maintaining, renewing, and extending suborbital facilities and capabilities with commercial operators. The lead would also monitor progress towards strategic objectives and advocate for enhanced suborbital activities, workforce development, and integration of suborbital activities within NASA. This would, in turn, lead to an increase in the number of space scientists, engineers, and system engineers with hands-on suborbital training. NASA could also use the suborbital flights as an integral part of on-the-job training and career development for engineers, experimental scientists, systems engineers, and project managers. Gradually, through a user-focused program, researchers, engineers, technologists, and educators would be able to conduct hands-on suborbital research. Low-cost, frequent flights would expand opportunities for hands-on learning, participatory research, and eventually personal experience in spaceflight that will be critical to developing technical and scientific competence. If the potential is demonstrated, one can envision that students would be able to take payloads from concept through operations via a NASA-sponsored academic program, thereby providing additional training options for the next generation of scientists. Maybe it will happen, maybe it won't, but, regardless of which scientists fly, they will all have to be medically certified and trained, so let's take a look at what those requirements may be.

3

Medical Qualification and Training

Although commercial astronauts will eventually work in orbit, in the near term, their work environment will be closer to home at suborbital altitudes of about 100 km. Now, you may think that medical certification and training for suborbital spaceflight will be a cinch; after all, with all the experience NASA has flying astronauts into space, flight surgeons should have a pretty good idea of what to look for and astronaut training should be fairly routine. In reality, that experience doesn't translate to the commercial suborbital space industry, and there are a number of reasons for this. First, the operational experience for manned suborbital space flight is very limited, consisting of just two Mercury-Redstone rocket flights in 1961 (Figure 3.1), two X-15 flights in 1963, a Soyuz launch abort in 1975, and three SpaceShipOne (SS1) flights in 2004. Second, if we were to apply the rigorous NASA medical certification and training standards to commercial astronauts, we'd find ourselves with a very small pool of candidates because not everyone has the résumé of a government-sponsored astronaut. Third, the stresses of suborbital flight are several orders of magnitude lower than those imposed by orbital flight so it stands to reason that medical certification and training reflect this. But what should the medical standards be and how much training should a commercial suborbital astronaut have? Should commercial suborbital astronauts just be required to meet the spaceflight participant (space tourist) medical standards? Should they do the same amount of training, or should they be required to meet a higher and more exacting standard?

Before discussing what medical standards the future corps of commercial suborbital astronauts will require, let's first define what a suborbital flight is. According to the Fédération Aéronautique Internationale (FAI), a suborbital space flight has to reach an altitude higher than 100 km[1] (62 miles, 328,000 ft) above sea

---

[1] This is the Kármán line, which is commonly used to define the boundary between Earth's atmosphere and space. The line was named after Theodore von Kármán (1881–1963), a Hungarian-American engineer and physicist who calculated that, at 100-km altitude, Earth's atmosphere becomes too thin for aeronautical purposes because any vehicle at this altitude would have to travel faster than orbital velocity to derive sufficient aerodynamic lift from the atmosphere to support itself.

44 Medical Qualification and Training

Figure 3.1 Close-up of Alan Shepard inside the Freedom 7 spacecraft (Mercury Capsule #7) as he communicates with flight controllers in the blockhouse at Pad 5 during the final moments of the countdown for the first US manned spaceflight. Courtesy: NASA

level, although the United States Air Force (USAF) and the Federal Aviation Administration (FAA) consider an altitude of 80 km (50 miles, 264,000 ft) as the altitude to qualify as space flight. While our experience in suborbital flight is limited, our knowledge of human performance below the international standard of space flight is much broader and can be taken into account when deciding what medical standards to use. Another important aspect we need to discuss is the environment during a suborbital commercial spaceflight and the potential medical risks that such a flight imposes. Since most people will be familiar with Virgin Galactic's SpaceShipTwo (SS2, Figure 3.2), we'll use their vehicle and flight profile as a baseline mission.

SPACESHIPTWO FLIGHT PROFILE

SS2 (Figure 3.2) will carry two pilots and up to six spaceflight participants. The cabin atmosphere will be pressurized to an altitude of 2,440 m and passengers will breathe re-circulated atmospheric air. The flight (Figure 3.3) begins with a horizontal take-off underneath the carrier aircraft WhiteKnightTwo (WK2) (Figure 3.4). Once WK2 reaches an altitude of 15,240 m, SS2 is launched. During the 70-second boost phase, passengers will experience acceleration loads of up to 3.8 G and speeds up to Mach 3 (maximum speed will be 4,180 km/h), 30 seconds after the rocket fires. The

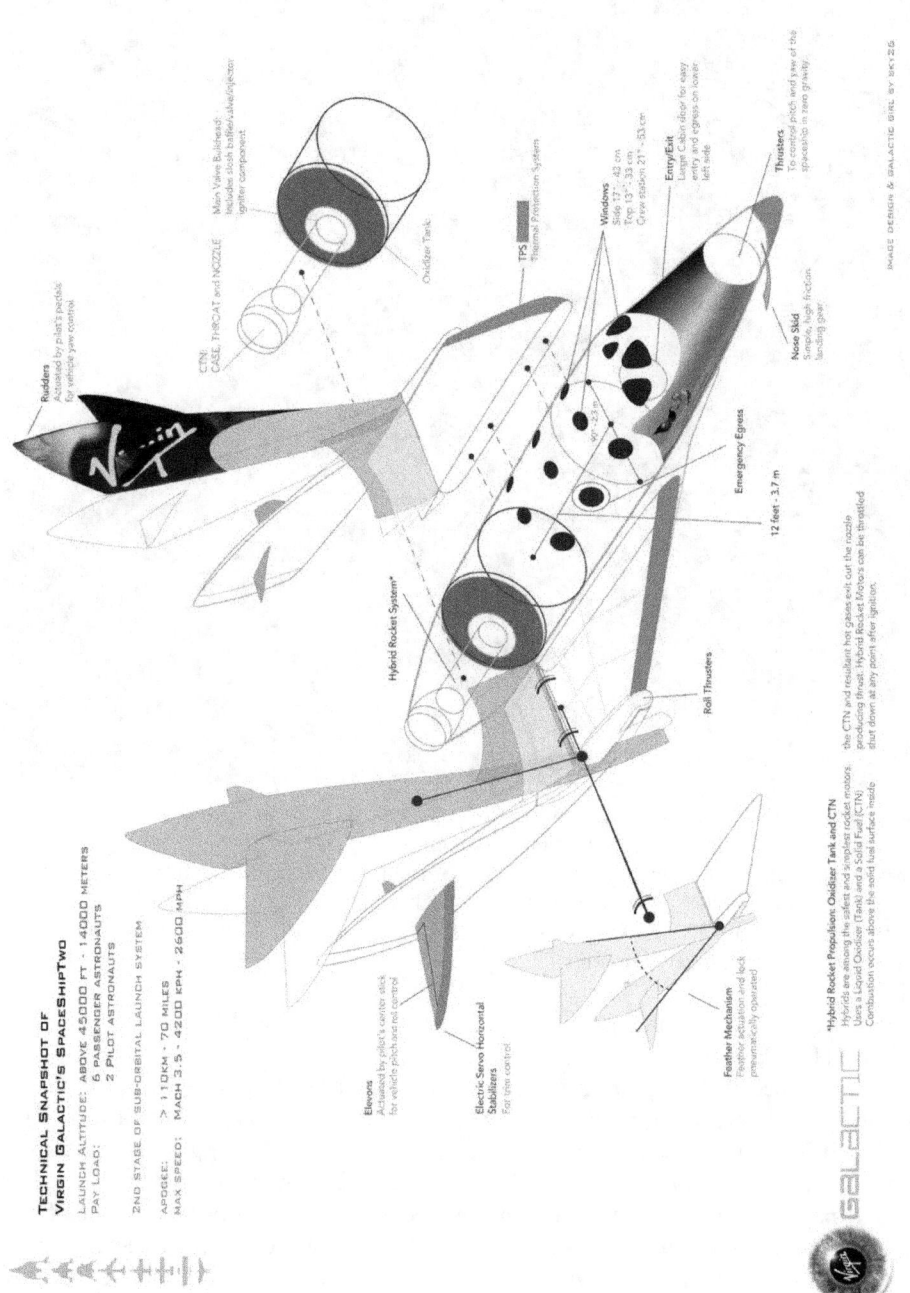

Figure 3.2 Technical description of SpaceShipTwo. Courtesy: The Spaceship Company

46 Medical Qualification and Training

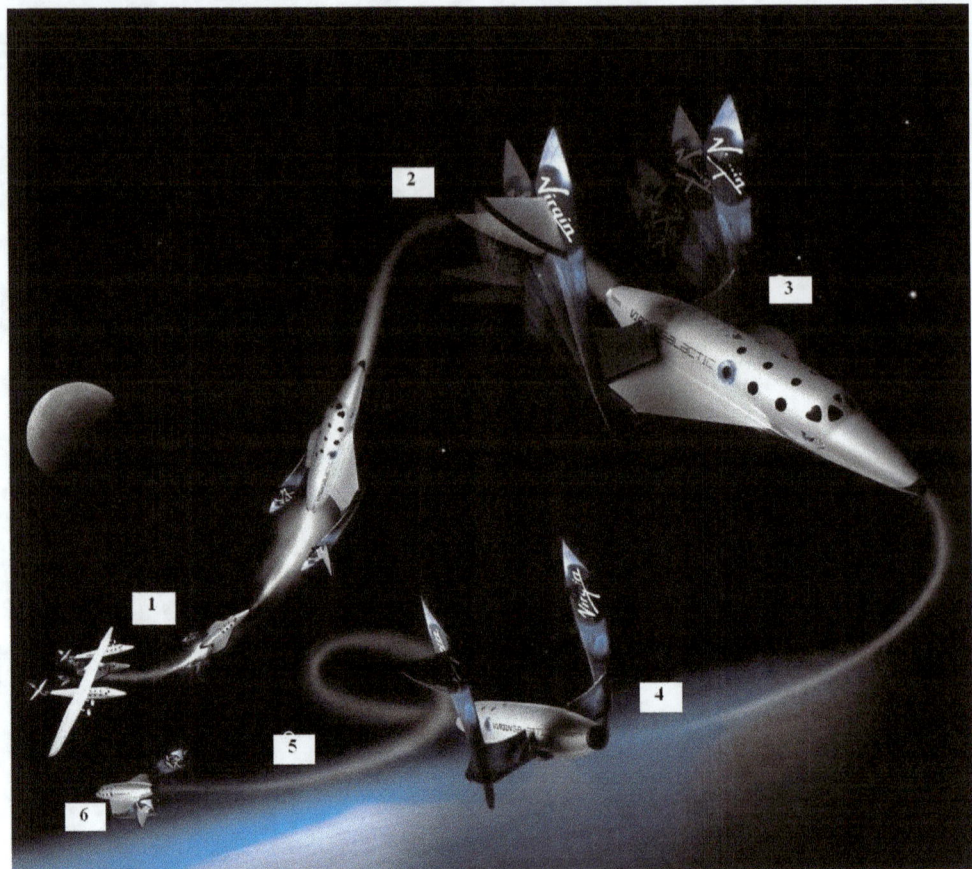

Figure 3.3 SpaceShipTwo's flight profile. (1) Air launch from 50,000 ft (15.5 km). Release from mother-ship and launch to Mach 4. (2) 328,000 ft (100 km). At this altitude, passengers become astronauts. (3) 361,000 ft. Virgin Galactic's maximum planned altitude. SpaceShipTwo feathers after rocket burn. (4) Re-entry in the feathered position. (5) 70,000 ft (21 km). De-feathered in glider mode. (6) Glide back to runway. Courtesy: The Spaceship Company

suborbital coast phase will last about four minutes and will reach a maximum altitude of 110 km.

During the deceleration phase, passengers will experience up to 6 G, but the seats will recline to convert most of the forces to +Gx (we'll get to an explanation of G-forces shortly). To increase stability and drag for entry, SS2's wings will rotate to a feather position (see details in Figure 3.2). Then, at 24,380 m, the 25-minute glide phase will begin with a return to an unpowered horizontal runway landing. Total flight duration will be about two-and-a-half hours. Now, of course, the radiation levels, noise environment, and vibration forces that will be experienced are still unknown, but we know enough about the space environment to be able to define the probable medical risks.

Figure 3.4 SpaceShipTwo in a captive flight underneath WhiteKnightTwo, during the runway dedication of Spaceport America in October 2010. Courtesy: The Spaceship Company

SUBORBITAL MEDICAL RISKS

If you watched the flights of SS1, then you don't need to be a flight surgeon to know that suborbital spaceflight exposes individuals to an environment that is much riskier than that experienced by those who fly on commercial airliners. Let's face it; if you have any pre-existing medical conditions, it's probably not a good idea to sign up for a flight that's going to make things worse. For the rest of the population, however, most of the medical issues for suborbital spaceflight are fairly straightforward, especially when compared to orbital spaceflight – just ask Richard Garriott[2] about the problems he faced when trying to be certified as a private space explorer! But, while the short duration of suborbital flights eliminates concerns for most of the medical problems associated with orbital flight such as bone loss, fluid shifts, and radiation exposure, that's not to say that suborbital flight isn't without some risks. After all, a critical aspect of suborbital spaceflight is

---

[2] For Richard Garriott to become a space tourist required some extra medical intervention, most notably the enucleation of a giant benign hemangioma, which had caused Garriott to suffer mission-excluding high blood pressure.

the rapid change from the high-G acceleration launch forces to zero-G weightlessness (the entry to which is preceded by pronounced deceleration) followed quickly by the high-G acceleration of entry. Even a benign flight profile like Virgin Galactic's is provocative enough to make the most ardent rollercoaster fan a little queasy and, from a medical standpoint, these transitions could lead to cardiovascular and neurovestibular effects that are currently unexplored. And there is just no way to completely simulate these forces or the suborbital flight environment pre-flight except for the actual experience of suborbital space flight; although the acceleration and deceleration forces can be simulated with centrifuge runs, the longest period of weightlessness that can be simulated with parabolic flight is only 25 seconds. This amount of time just isn't long enough for all the neurovestibular and cardiovascular reflex changes to occur. More importantly for flight surgeons, the environment of acceleration-to-suborbital zero-G-to-acceleration has never been simulated (except for the brief $+1.8$ G to 0 G to $+1.8$ G experienced in parabolic flights) and is quite different from the G's pulled in orbital spaceflight. The bottom line is that the operational experience of manned suborbital space flight is extremely limited, so, before we decide what medical standards our future corps of suborbital commercial astronauts will require, let's take a look at the medical effects experienced during a typical flight.

Acceleration

We'll start by discussing the G-forces of the flight profile. Seen through the eyes of the flight surgeon, the most worrying aspects of a suborbital flight profile are the launch acceleration and entry deceleration, especially when the acceleration exposure is in the head-to-foot ("eyeballs down" or $+Gz$) direction (Figure 3.5). That's because Gz acceleration can cause all sorts of neurovestibular, cardiovascular, and musculoskeletal problems [1, 2]. Exposure to $+Gz$ can also have an impact on pulmonary function proportionally to its applied force magnitude – at the lower end of the G-scale, say $+2-3$ G, commercial spaceflight astronauts may experience difficulty breathing, while at the other end of the G-load spectrum, say $+5-6$ G, astronauts may suffer airway closure and even atelectasis.[3]

To avoid these problems, spacecraft designers try to limit the launch and re-entry acceleration forces by ensuring most of the acceleration is in the $+Gx$ direction (eyeballs in). That's because an individual is more tolerant to $+Gx$ acceleration and, with the heart and brain located at approximately the same level within the acceleration field, there is less risk of gravity-induced loss of consciousness (G-LOC) or almost loss of consciousness (A-LOC). Acceleration stress is perhaps one of

---

[3] During typical $+Gz$ acceleration, blood pools at the base of the lungs, causing edema and constriction of the distal airways, reducing local ventilation. The acceleration also displaces lung tissue, stretching and expanding the alveoli (air sacs in the lungs). This displacement can occasionally be sufficient to cause pulmonary atelectasis due to the rapid absorption of oxygen from closed-off alveoli near the base of the lung. Usually, these symptoms diminish and disappear following several deep breaths.

Suborbital medical risks 49

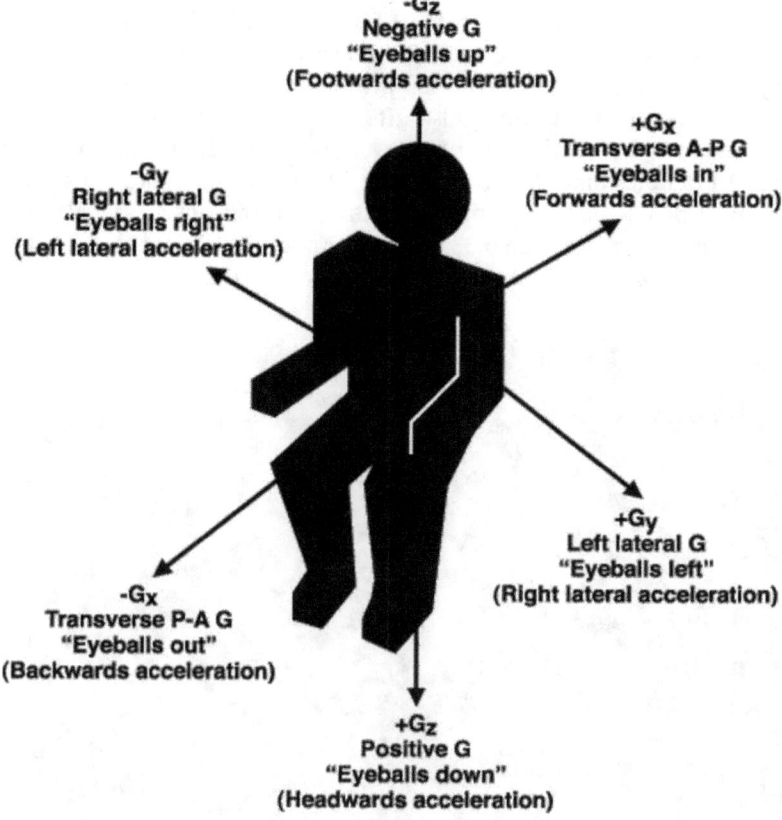

Figure 3.5 The different types of G. Courtesy: NASA

the issues that most worry the flight surgeon because it is dysrhythmogenic, which means the heart's rate, rhythm, and conduction can be affected. In fact, high G-forces or particularly long exposures to acceleration could potentially increase the frequency of a heart problem known as a dysrhythmia. It is for this reason that spaceflight accelerations have mostly been designed to be in the +Gx-axis – that is, until the Shuttle came along. In the early days of manned spaceflight, the direction of acceleration was even more important than it is today because of the sheer magnitude of the acceleration. For example, the Mercury, Gemini, and Apollo flights had launch accelerations of 4.5–6.5 +Gx for six minutes and anywhere from 6 to 11 +Gx during entry (which was why astronauts received 45 hr of +Gx centrifuge training, with some runs going up to 18 +Gx!). By comparison, the now-retired Shuttle had a maximum of 3.0 +Gx during the 8.5-minute launch and 1.2 +Gz (briefly 2.0 +Gz during turns) for 17 minutes during entry. Fortunately for the new breed of commercial astronauts, the acceleration forces imposed by most of the current crop of space vehicles should be reasonably comfortable for most people, although there will be some who will do better than others.

50 **Medical Qualification and Training**

You see, an individual's tolerance to +Gz acceleration is dependent on their height and weight, certain physiological characteristics, and the type of acceleration profile; physical conditioning, hydration, previous and recent exposure to +Gz forces, and recent centrifuge training also affect your response to acceleration. This is important because the maximum +Gz level, exposure duration, and the rate of onset of the +Gz determine the risk of injury to your heart and musculoskeletal system. The most problematic type of acceleration is rapid-onset rate (ROR), which is defined as increases greater than 0.33 G per second [1, 2]. ROR tolerance limits are approximately 1 +Gz lower than gradual-onset rate (GOR) tolerances because they

Figure 3.6 The author being fitted for G-pants before his Hawk flight. Courtesy: Author

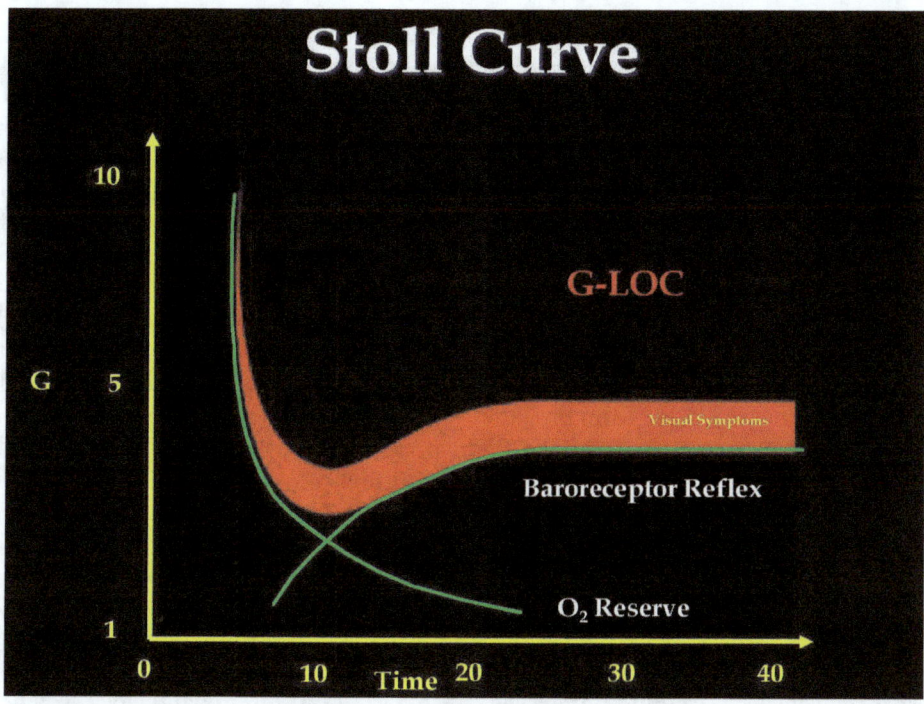

Figure 3.7 The Stoll curve describes the five-second oxygen reserve during which no visual symptoms occur. The dip in the curve illustrates the problem caused by the lag in physiological compensatory mechanisms, especially with high onset rates of G. Courtesy: NASA

exceed the ability of the cardiovascular system to fully respond to preserve adequate central nervous system (CNS) blood flow; basically, if your brain doesn't get enough blood, it will shut down. RORs can also result in the dreaded G-LOC without any of the usual visual warning symptoms such as tunnel vision, gray-out, or black-out. To prevent this happening when they're performing aerobatic maneuvers, fighter pilots wear anti-G suits (Figure 3.6), which increase their G-tolerance to +Gz by up to 1.5 +Gz. Another way fighter pilots can increase their G-tolerance is to practice the anti-G straining maneuver (AGSM), which can increase tolerance to +Gz by as much as 3 +Gz. However, performing the AGSM is tiring and is generally used only for a short period of time.

Over the years, centrifuge data have allowed scientists to develop a model of +Gz tolerance limits that incorporate +Gz magnitude, duration, and rate of onset (generally, with no protection, most healthy people can tolerate up to 4 +Gz acceleration for ROR profiles and up to 4.5 +Gz with GOR profiles). The model is called the Stoll curve (Figure 3.7) after the investigator who first described it in 1956.

Okay, so we've discussed +Gz tolerance, but what about −Gz and the transition from one type of G to another? Well, this is perhaps where most of the problems occur because transition from −Gz to +Gz can cause a profound drop in cerebral blood pressure and that's bad news for the cardiovascular system because it can take quite a while before the body compensates. In fact, prior exposure to −Gz followed immediately by +Gz can cause the deadly "push–pull effect". Usually, this "push–pull effect" occurs in combat engagements and has been implicated in several combat training fatalities (it's also been identified as a possible cause of 30% of G-LOC events). Even now, with a wealth of G-data available, there is a knowledge gap in the complete understanding of this issue and no countermeasures have been developed. It's unclear whether a "push–pull effect" will occur in the transition from microgravity to entry deceleration, but it has been experienced in parabolic flight [2] and there are some who are concerned that it could occur in suborbital flights. That's because the "push–pull effect" is prolonged by increasing the duration of the prior −Gz exposure [2]. Normally, the −Gz exposure is only several seconds in combat flight, whereas, in parabolic flight profiles, the exposure is 20–30 seconds. But what about after four minutes of suborbital flight? We just don't know whether four minutes of microgravity would provoke the same response or a further deterioration in the +Gz tolerance. Maybe that's a research project that can be tested on future commercial suborbital astronauts?

Now, if you're planning to fly as a passenger on board one of the new crop of space vehicles and you're worried about how you might be affected by G, you should know that the acceleration envelope recommended by the IAA for commercial aerospace vehicles [3] should not exceed 3 +Gz (2 −Gz), ±6 Gx and ±1 Gy. These levels, if experienced as GORs, should be well tolerated by unprotected, healthy individuals. Let's take Virgin Galactic's vehicle as an example. During SS2's rocket engine boost, acceleration may be as high as 3.8 +Gx followed by a brief spike up to 4.0 +Gz as the vehicle rotates to a nose-high attitude. On re-entry, 6 G will be felt mainly in the +Gz-axis by the pilots but, thanks to SS2's tilt-back seating, most of the acceleration during entry will be in the Gx-axis for the passengers. Duration of these G-forces is expected to be about 70 seconds during launch and about 30 seconds during re-entry. Although SS2's acceleration onset rate has yet to be defined, it isn't expected to be an ROR.

Microgravity effects

The physiological changes resulting from exposure to microgravity depend upon the duration of the exposure and, like acceleration, can vary in magnitude from individual to individual. While suborbital microgravity exposure will only last for four or five minutes, it's possible that inexperienced, non-adapted, or highly sensitive passengers may experience a variety of neurovestibular or cardiovascular symptoms. Although no proof exists, parabolic flight experience might be a way to alleviate suborbital flight symptoms by providing weightless experience, which is why Astronauts for Hire (A4H) includes zero-G training as part of their astronaut training. Once again, there are few operational data about microgravity effects in suborbital flight (limited to Mercury-Redstone, the X-15 flights – Figure 3.8 – and

Figure 3.8 Neil Armstrong in the X-15 #1 cockpit. Photo ECN-89, taken in 1961. Courtesy: US Navy

the 2004 SS1 flights), so this may well be another research opportunity that commercial suborbital astronauts find themselves studying.

Cardiovascular effects

A common cardiovascular effect that was observed in Shuttle astronauts while they were lying down awaiting launch was an increase in central venous pressure (CVP). This was followed by a decrease in CVP to below normal levels almost as soon as the astronauts reached space [4]. Part of the reason for the increase in CVP has been attributed to the slightly head-down pre-launch position and also due to the astronauts being dehydrated during the pre-launch period. The decrease in CVP has been attributed to a reduction in intra-thoracic pressure and a loss of gravitational compression [5]; in orbit, due to the absence of gravity, body fluids rush to the head, giving astronauts a sensation of head fullness, which feels a little like having the flu [5]. Most of these effects won't be a problem during suborbital flight simply because the physiological changes take time to develop in microgravity. For example, the SS2 passengers and crew won't be spending much time in a pre-launch horizontal position (the Shuttle astronauts spent hours lying down), so their body fluids won't shift very much.

Neurovestibular effects

In common with the fluid shifts, the neurovestibular effects won't be a significant factor for most passengers in suborbital flight because these effects take time to manifest themselves. After orbital flight, as we'll see later in this chapter, astronauts suffer an altered ability to sense tilt and roll, defects in postural stability, impaired gaze control, and changes in sensory integration [3]; basically, they're completely discombobulated! While suborbital passengers probably won't be affected to the degree that Shuttle astronauts were, simply because the changes are dependent on the duration of weightlessness, there have been neurovestibular alterations observed in even short exposures to zero-G in susceptible individuals. For example, somatogravic illusions with spatial disorientation have been reported on several of the high-altitude X-15 flights. Now, you may think why not test this in a simulator, but the problem is that the flight profile of rapid launch acceleration followed by zero-G followed by re-entry deceleration can't be tested in continuity. Another problem is that many commercial astronauts will be flying these suborbital profiles repeatedly and maybe even on a daily basis and there isn't any experience that shows whether repeated and frequent suborbital profile exposures will be adaptive or maladaptive to neurovestibular function. Obviously, more experience in suborbital flight is needed and it will be the job of the new corps of commercial suborbital astronauts to determine whether this is even an issue. One way this will be done will be by conducting post-flight medical debriefs and data collection in the form of biomedical monitoring – an initiative that A4H is already working on.

X-15 neurovestibular experience

Although we don't have much suborbital flight experience, it's worth taking a look at the experience of those pilots who flew similar flights back in the 1950s and 1960s. The most notable suborbital flights were flown by the X-15 (Figure 3.9), three of which were built for a program that comprised 199 test flights; the first was flown on June 8th, 1959, and the last one on October 24th, 1968.

Among the 12 pilots of the X-15 program were future NASA astronauts Neil Armstrong and Joe Engle. During the program, 13 of the flights (by eight pilots) met the USAF space flight criteria by exceeding an altitude of 80 km, thus qualifying the pilots for astronaut status. Of all the X-15 missions, only two flights[4] (both piloted by Joe Walker) qualified as spaceflights according to the FAI definition. Although biomedical data weren't measured on the X-15 flights, pre-flight and post-flight flight surgeon examinations were performed and nothing unusual was reported.

---

[4] Flight 90 on July 19th, 1963, reached 105.9 km. Flight 91 on August 22nd, 1963, reached 107.8 km.

Figure 3.9 This US Air Force photo shows the X-15 Ship #3 (56-6672) in flight over the desert in the 1960s. Ship #3 made 65 flights during the program, attaining a top speed of Mach 5.65 and a maximum altitude of 354,200 ft. Only 10 of the 12 X-15 pilots flew Ship #3, and only eight of them earned their astronaut wings: Robert White, Joseph Walker, Robert Rushworth, John "Jack" McKay, Joseph Engle, William "Pete" Knight, William Dana, and Michael Adams. Courtesy: US Air Force

Also, pilot performance wasn't impaired by launch acceleration followed by zero-G followed by re-entry, although the G-forces imposed on the pilot during the boost phase of the flight often resulted in severe vertigo. In fact, vertigo was a serious problem throughout the X-15 program and was blamed for the death of Major Mike Adams (Figure 3.10).

Major Adams joined the X-15 program in October 1966. His seventh X-15 flight took place on November 15th, 1967, in the number three aircraft. At 10:30 in the morning the fateful day, Major Adams's X-15 dropped away from the NB-52B at 45,000 ft. At 10:33, he reached a peak altitude of 266,000 ft. On reaching its maximum altitude, the X-15's heading was off by 15 degrees and, as Adams descended, the aircraft began drifting to the right. Then, at 230,000 ft, encountering rapidly increasing dynamic pressures, the X-15 entered a Mach 5 spin.

In the NASA 1 control room, controllers advised Adams that he was "a little bit high", but in "real good shape". Adams replied that the aircraft "seems squirrelly". At 10:34, a troubling call came: "I'm in a spin." With no heading information, controllers saw only large and very slow pitching and rolling motions and thought Adams was overstating the case. But Adams radioed again: "I'm in a spin." A NASA test pilot sitting in the control room said quietly: "That boy's in trouble." Unfortunately, there was no spin recovery technique for the X-15, and engineers

Figure 3.10 Michael J. Adams with X-15. Photo ECN-1651, taken on March 22nd, 1967. Air Force test pilot Maj. Michael J. Adams stands beside X-15 Ship #1. Adams was selected for the X-15 program in 1966 and made his first flight on October 6th, 1966. On November 15th, 1967, Adams made his seventh and final X-15 flight. The X-15 launched from the B-52 but, during the ascent, an electrical problem affected the X-15's control system. The aircraft crashed north-west of Cuddeback Lake, California, causing the death of Adams. Adams was the only pilot lost in the 199-flight X-15 program. Courtesy: US Air Force

knew next to nothing about the aircraft's supersonic spin tendencies. The chase pilots, realizing the X-15 would never make Rogers Dry Lake, turned on the afterburners and raced for the emergency lakes. Somehow, the aircraft recovered from the spin at 118,000 ft and went into an inverted Mach 4.7 dive. Adams now had a good chance of rolling upright, pulling out, and setting up a landing. But then the X-15 began a rapid pitching motion, still in its dive at 160,000 ft per minute. Dynamic pressure increased intolerably and, as the aircraft neared 65,000 ft, it was diving at Mach 3.93 and experiencing over 15 G vertically, both positive and negative, and 8 G laterally. The aircraft broke up 10 minutes 35 seconds after launch.

NASA and the Air Force convened an accident board, which concluded that

Adams[5] suffered severe vertigo during climb-out, which caused spatial disorientation. Small heading deviations caused by a degraded flight control system were made worse by incorrect pilot inputs at an altitude of over 20 km. The board concluded that Adams had misinterpreted a roll indication for a slide-slip indication and made control inputs in the wrong direction. As a result of the accident investigation, it was recommended that all future X-15 pilots be medically screened for labyrinth (vertigo) sensitivity. The board also recommended that a telemetered heading indicator be installed in the control room, visible to the flight controller.

Space motion sickness

More than 70% of first-time astronauts flying on orbital spaceflights suffer from space motion sickness (SMS). The syndrome, thought to be due to a sensory conflict [6, 7] between visual, vestibular, and proprioceptive stimuli, has been a problem as long as there have been astronauts. As well as being uncomfortable for the astronauts, SMS is a major headache for space agencies because it's impossible to predict who will be affected and when they will be affected; symptoms typically occur within the first 24 hr, but some astronauts have reported symptoms – dizziness, pallor, sweating, severe nausea, and vomiting are the most common – immediately after main engine cut-off. Vomiting, which can be especially messy in zero-G, can crescendo suddenly without any warning symptoms. In a multi-passenger vehicle, one passenger becoming nauseated can potentially trigger nausea in the other vehicle occupants – just imagine trying to do your work as a commercial astronaut while barfing into a vomit bag or trying to dodge balls of stomach contents as they are ejected from your fellow suborbital workers! Anti-motion sickness medications could be used, but then performance would be affected (Space Shuttle flight crew members were not allowed to take prophylactic medications for SMS for precisely this reason). The risk of nausea in zero-G can be reduced if provocative motions, especially of the head, are avoided. You see, head movements generate conflicts between the semicircular canals and the otoliths, with pitch movements being the most provocative [8].

Unfortunately, there is just no way to protect you against the possibility of being sick. Parabolic flight adaptation and experience in high-performance jet aircraft don't work. Neither do rotating chairs nor centrifuge training. The only preflight training that has been shown to be effective is the use of training aids that duplicate the sensory conflict that occurs in parabolic flight; this type of pre-flight adaptation training is an attempt to make passengers "dual-adapted". An example of the training aids used in this effort to duplicate sensory conflict is the device for orientation and motion environment (DOME), which is a spherical virtual reality simulator. During suborbital flights, the risk of SMS will be reduced if flight

---

[5] Mike Adams was posthumously awarded Astronaut Wings for his last flight in the X-15-3, which attained an altitude of 266,000 ft. In 1991, Adams's name was added to the Astronaut Memorial at the Kennedy Space Center in Florida.

58 Medical Qualification and Training

crewmembers remain strapped into their seats during the flight and limit head movements, although this might prove difficult when attending to a payload.

Emergency egress capability

One of the most important medical and training assumptions is that crewmembers will be capable of performing an emergency egress without assistance. Obviously, performing an emergency egress will depend on the type of vehicle you're flying, as some vehicles, such as SS2, don't have such a capability, but that's another story. Back in the days when the US had its own manned spaceflight capability, up to 15% of Shuttle astronauts were judged to be too impaired post landing to perform an unaided egress [8]. The reason the Shuttle astronauts were so discombobulated was due to a combination of entry motion sickness, post-flight neurological dysfunction, and post-flight orthostatic intolerance. Fortunately, as these problems are dependent on the duration time of microgravity, they shouldn't be a frequent concern after suborbital flights, but there are always susceptible individuals who may have problems; it's one of the reasons why A4H includes emergency egress training (Figure 3.11) as part of its astronaut training.

Figure 3.11 A4H members conducting emergency egress training at Survival Systems, July 2011. Courtesy: A4H

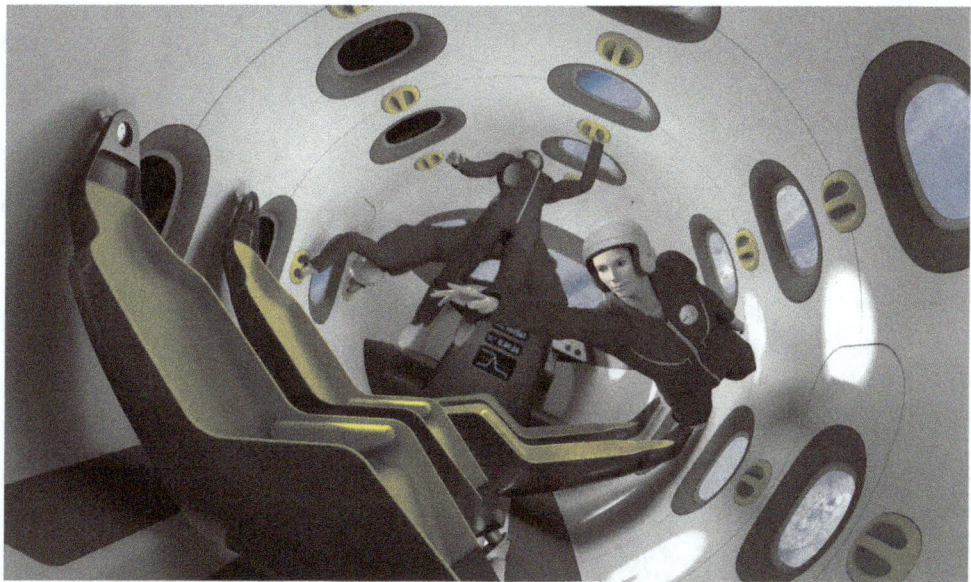

Figure 3.12 The interior of SpaceShipTwo. Courtesy: The Spaceship Company

Environmental medical issues

You may think that the cabin environment is a relatively minor operational concern, but a malfunction of the cabin heating, air circulation, and/or cooling systems could result in an uncomfortable cabin environment that affects cognitive and psychomotor performance. Also, SMS is known to be exacerbated by over-heating, and then there's the issue of cabin pressure. In common with cabin temperature and humidity (in most orbital space flight vehicles, the cabin temperature is typically 21–26°C with a relative humidity of 30–40%), cabin pressure will vary, depending upon the design of the space vehicle. In current orbital space vehicles, the cabin pressure is maintained at a sea-level pressure of 14.7 psi (101 kPa), allowing for a shirt-sleeved environment (Figure 3.12), which some people think isn't a good idea, and here's why.

You see, suborbital vehicles will operate at such high altitudes that there is a potential risk of a rapid or explosive decompression, either of which could result in hypoxia or even death due to hypoxia or ebullism. In case you're wondering how badly such an event might be, here's what *The USAF Flight Surgeon's Guide* has to say about some of the effects due to mechanical expansion of gases during rapid decompression:

> "One of the potential dangers during a rapid decompression is the expansion of gases within body cavities. The abdominal distress during rapid decompression is usually no more severe than that which might occur during slower decompression. Nevertheless, abdominal distention, when it does occur, may have several important effects. The diaphragm is displaced upward by the

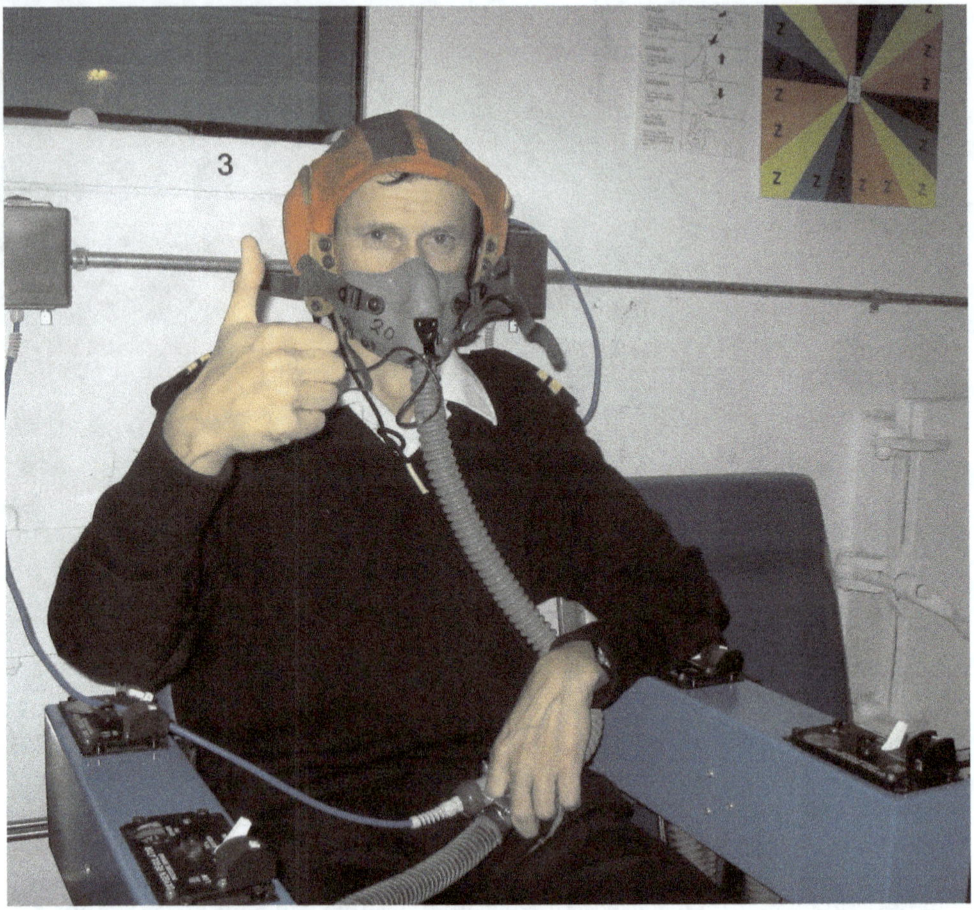

Figure 3.13 The author during a high-altitude indoctrination run. Courtesy: Author

expansion of trapped gas in the stomach, which can retard respiratory movements. Distention of these abdominal organs may also stimulate the abdominal branches of the vagus nerve, resulting in cardiovascular depression, and if severe enough, cause a reduction in blood pressure, unconsciousness, and shock."

Sounds painful, doesn't it? Well, rapid decompression is probably more survivable than explosive decompression – an event that can cause your blood to literally boil and your body to swell to twice its normal size. Here's what Dr Tamarack R. Czarnik, a specialist in aerospace medicine, has to say about the medical effects of an explosive decompression:

"Damage to the lungs in rapid or explosive decompression occurs primarily due to pulmonary overpressure, the tremendous pressure differential inside versus outside the lungs. 80 mm Hg is enough to cause pulmonary tears and alveolar rupture; pulmonary hemorrhaging, ranging from petechiae to free blood is also seen. Emphysematous changes are seen especially in the upper lungs, while atelectasis and edema predominate in the lower lungs. When we get to the patient, the lungs will be a bloody, ruptured mess. Though these are the most life-threatening changes seen in ebullism, subcutaneous swelling is also seen, due to creation of water vapor under the skin. This can rapidly distend the body to twice its normal volume. Our patient will look no better than he feels, though this means little in terms of survival."

Of course, the easy way to avoid these nasty hypobaric events would be to use a pressure suit, but pressure suits are expensive, they weigh a lot (the Shuttle pressure suit weighed over 20 kg), they make you sweat, and they decrease performance. It is noted that a pressure suit was not used on the SS1 flights. Once again, it comes down to training as a means to mitigate the risks, which is why most operators include high-altitude indoctrination (Figure 3.13) as a means of familiarizing their passengers with the hazards of hypoxia and depressurization.

Radiation

When we talk about harmful radiation in space, we're generally talking about ionizing radiation. This type of radiation consists of subatomic particles that can interact with biological tissues and destroy DNA strands, causing genetic damage that can, in turn, lead to dangerous mutations. The sources of ionizing radiation in space are galactic cosmic radiation (GCR), solar radiation, solar flares, and the trapped radiation from the Van Allen belts. GCR originates outside of the Solar System and consists of hydrogen nuclei protons (87%), helium nuclei alpha particles (12%), and damaging high-energy heavy nuclei such as iron (1%), while solar cosmic radiation (SCR) comprises proton–electron plasma ejected from the surface of the Sun. Completing the radiation cocktail are solar flares (Figure 3.14), which are magnetic disturbances on the Sun's surface generating electromagnetic radiation.

So, how much radiation is too much? Well, the dose standard for radiation-exposed workers is 20 mSieverts (mSv) per year (averaged over five years); exposure to this level over a work life of 40 years results in an excess lifetime fatal cancer risk of 3.2%. By comparison, orbital spaceflight results in a variable radiation dose exposure, dependent on orbital altitude and solar activity, and ranges from 0.01 to 0.1 Sv/month. NASA astronauts have career exposure limits based upon a maximum of 3% excess lifetime cancer mortality [9]; these limits are recommended by the National Council on Radiation Protection (NCRP). Planned exposures to radiation such as during an EVA when an astronaut isn't shielded by the spacecraft are kept as low as reasonably achievable (the ALARA principle).

Radiation levels at a suborbital flight altitude would be similar to high-altitude Concorde flights, which means commercial suborbital astronauts should receive less than 15 microSv/hr [10] during their flight. It doesn't sound a lot but, remember, these astronauts will be flying multiple missions, so what are the recommended occupational exposure limits? Well, the International Commission on Radiological

62 Medical Qualification and Training

Figure 3.14 Solar flare activity can increase the risk of radiation exposure. Courtesy: NOAA

Protection (ICRP) says that, for commercial aircrews, such as on Concorde, the occupational limit is 20 mSv/year, averaged over five years, with a maximum in any one year of 50 mSv. Compare this to the ICRP recommendation for the general public that exposure be less than 1 mSv/year.

Unlike government-employed astronauts, who are exposed to ionizing radiation for months on end, commercial astronauts shouldn't be too worried because their flights will be so short and the operators will have the advantage of controlling the launch in the event of unfavorable atmospheric conditions.

Noise

Launching a vehicle into suborbital space requires powerful thrust, which happens to be very noisy. This noise is transmitted through the whole vehicle and, because the vehicle is an enclosed space, the noise is reflected multiple times off the walls, floor, and ceiling. Although the noise levels are relatively short, the magnitude can be quite intense – so intense that physiological effects such as reduced visual acuity, vertigo, nausea, disorientation, and ear pain may be experienced. Loud noise can also interfere with normal speech, making it difficult to communicate, which, in turn, can affect team interaction and result in task errors, especially in those tasks requiring multiple information sources, information processing, and vigilance [11]. Noise levels in the crew compartment during a Shuttle launch reached almost 120 dB (equivalent to an amplified rock concert in front of the speakers). Because of this assault on your hearing, auditory protection will definitely be required during a suborbital launch.

Vibration

As well as all that noise, the power being unleashed to power the vehicle in suborbital flight will also generate significant vibration (check out the in-cabin videos of the SS1 flights during both ascent and entry and you'll see what I mean); think about the vibration you feel when an aircraft takes off, multiply that by about 10 orders of magnitude, and you have some idea of what to expect. While vibration won't be more than a temporary inconvenience for the tourists, for commercial astronauts tasked with flying payloads it could be a problem. That's because vibration can cause manual tracking errors and can interfere with your ability to read displays; this might be a problem for someone tasked with keeping an eye on an experiment from launch through to re-entry.

Astronauts for Hire suborbital medical standards

The A4H Training Committee decided that the current Federal Aviation Administration (FAA) Class III medical standards were appropriate for suborbital payload specialists. The committee also decided that a waiver system similar to that which exists for flying organizations be adopted for the A4H medical certification. The purpose of the medical examination is to ensure the suborbital payload specialist meets medical standards during the period of certification and to detect common illnesses that may cause sudden incapacitation. In addition to the medical standards, the A4H training committee decided to implement long-term health and fitness monitoring, which will include annual indirect oxygen uptake assessment, body-fat composition, and strength assessment.

The A4H Training Committee considered only suborbital flight medical requirements because this will be a reality in the very near future. It's likely that orbital flights will be offered to commercial payload specialists within the next five years and, because orbital flight will impose greater stresses than suborbital flight, it will be addressed separately. Based on an evaluation of the medical risks in the suborbital realm, the A4H Training Committee proposed the following recommendations for their suborbital payload specialists:

- a FAA 3rd-class medical certificate;
- pre-flight medical evaluations within one week of flight to reduce risk and liability if any unpredicted medical issues occur;
- post-flight medical debriefs with data collection;
- the development of an independent longitudinal health monitoring database and periodic re-evaluation of medical standards with recommendations for changes to respond to medical issues that may be discovered;

- passive ionizing radiation dosimeters worn by each crewmember;
- auditory protection in the headset for each crewmember;
- generic and vehicle-specific emergency egress training to be conducted by each crewmember;
- theoretical and practical physiological environmental training (altitude chamber) to ensure flight crew recognition of signs and symptoms associated with decompression, such as rapid and explosive decompression;
- recent centrifuge (within 36 months of flight) training that exposes crewmembers to acceleration forces in excess of 3.5 +Gz;
- recent parabolic flight training (within 36 months of flight) to develop tolerance with adaptation to the changing gravitational fields.

TRAINING FOR COMMERCIAL SUBORBITAL SPACEFLIGHT

"I am living in fear of the move to develop standards for crew and passenger training in this industry. I think it's a mistake at this stage in the development of the industry."

Jeff Greason, CEO of XCOR Aerospace

Astronauts employed by government space agencies receive extensive training after their selection and, when picked for a mission, they undergo even more training, which can last a year or more. For suborbital spaceflight, which will last only a matter of minutes, there's no need for so much training, but exactly how much training is appropriate? It's a question that has yet to be answered because the industry has not yet standardized astronaut training and, for the moment, it doesn't want to. Even the FAA has provided only vague guidelines[6] as to what suborbital flight training should include. Of course, there are a number of training options available for spaceflight participants and even those hoping to be employed as commercial astronauts. For example, the National AeroSpace Training and Research (NASTAR) Center (Figure 3.15) was used by Virgin Galactic to train some of its spaceflight participants; the center also delivers a Suborbital Scientist Training Program (SSTP) for those interested in being paid to fly on a suborbital flight.

[6] According to the FAA, a suborbital operator must: "(1) Ensure that any crew-training device used to meet the training requirements realistically represents the vehicle's configuration and mission, or (2) Inform the crew member being trained of the differences between the two. (c) Maintenance of training records. An operator must continually update the crew training to ensure that it incorporates lessons learned from training and operational missions. An operator must – (1) Track each revision and update in writing; and (2) Document the completed training for each crew member and maintain the documentation for each active crew member. (d) Current qualifications and training. An operator must establish a recurrent training schedule and ensure that all crew qualifications and training required by § 460.5 are current before launch and reentry." For those interested in what a suborbital training schedule might include, a sample program is presented in Appendix II.

Figure 3.15 Brienna Henwood introduces the NASTAR Center. Courtesy: A4H

NASTAR

The NASTAR Center is a non-government, world-class aerospace training facility that supports the training, research, and educational needs of the aerospace industry. Established in 2006, NASTAR began as a product showcase, engineering development, and test center for its owners, the Environmental Tectonics Corporation (ETC), but soon became recognized within the aerospace training industry for its unique approach and sophisticated interactive flight training technology. The center contains all sorts of space training equipment, ranging from high-fidelity simulators to a multi-axis centrifuge (Figure 3.16) – ATFS-400 Phoenix, which you may recognize if you've seen news articles about Virgin Galactic's training.

In June 2009, shortly after Virgin Galactic started using NASTAR for their spaceflight participant training, Alan Stern and Dan Durda of the Southwest Research Institute (SwRI) paid the center a visit with the aim of establishing a training program for suborbital scientists. In August 2009, SwRI and NASTAR established the Suborbital Scientist Training (SST) course, a two-day program designed to qualify individuals for the physiological rigors of suborbital human spaceflight and to prepare them for suborbital Research and Education Missions (REM):

Figure 3.16 NASTAR's AFTS-400 centrifuge. Courtesy: NASTAR

"NASTAR is the leader in private space flight training. Having trained for and flown high-performance F-18s and high-altitude reconnaissance jets like the WB-57, I could see that NASTAR was uniquely positioned to support this program with its extensive, modern facilities and their application to suborbital research in space by scientists, engineers, educators and even students."
  Alan Stern, associate vice president of the SwRI Space Science and Engineering Division, commenting on NASTAR's Suborbital Scientist Training program

On the first day of the SSTP, students are given a tour of the facility during which they are shown the ejection seat simulators, the yaw, pitch and roll confusion generators (the real name is slightly different), the hypobaric chamber, and, of course, the high-tech centrifuge, complete with spaceflight simulator pod. Then, they're fitted for their flight suits and instructed on the basics of aerospace physiology. In the afternoon, the budding suborbital scientists complete their high-altitude indoctrination inside the hypobaric chamber (Figure 3.17) at 18,000 ft for 20 minutes while attempting to solve problems on a worksheet to test their reaction to hypoxia.

Day Two starts with the preparation for the G-tolerance test, which includes teaching the students the anti-G straining maneuver (AGSM). Performing the AGSM helps combat G-forces in the Gz-axis. After practicing their AGSM

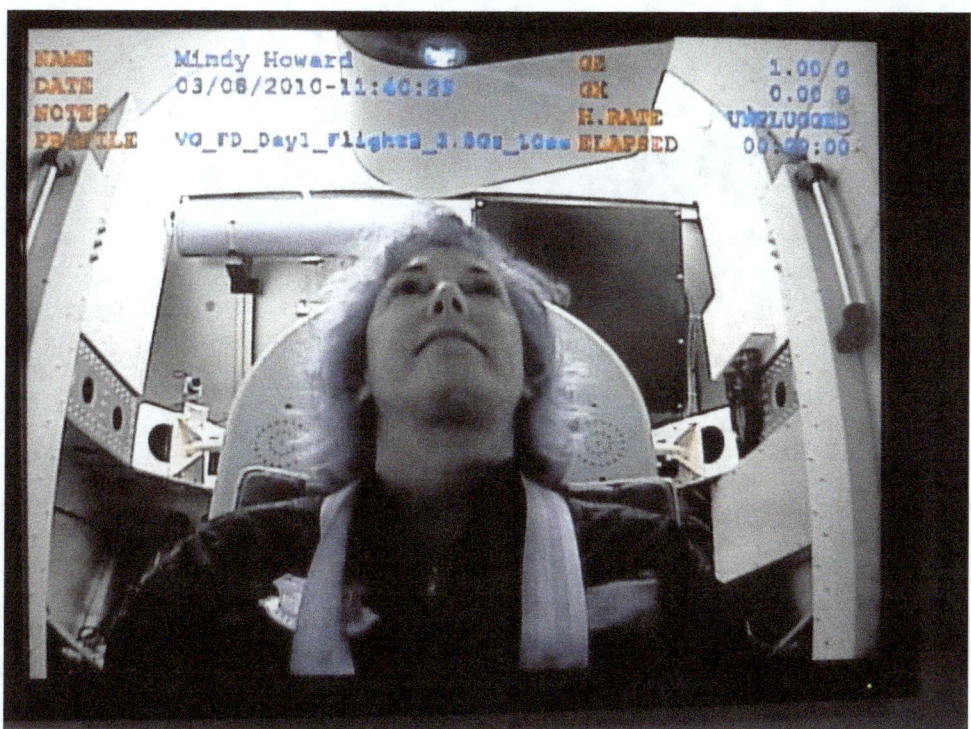

Figure 3.17 A4H member, Mindy Howard, takes a ride in the NASTAR centrifuge. Courtesy: Mindy Howard/NASTAR/A4H

technique, the students perform four centrifuge runs (NASTAR likes to call the runs "Flights"): two Gz at 2 and 3.5 G and two Gx at 3 and 6 G. After surviving the centrifuge, students take part in a distraction exercise, which takes place in a room with a black interior and set up with a Virgin Galactic SS2-sized cabin area. During the exercise, a projector plays space images as if viewed out of one of SS2's windows. At the front of the simulated cabin are six boxes representing experiments. The task for the students is to open their assigned boxes and perform the experiment within five minutes, which happens to be the period of zero-gravity time. The experiments usually include tasks such as matching patches around the cabin with patches in the box to finding odd items in a picture. Students typically find the whole exercise to be a discombobulating experience the first time round, although, by the second attempt, they get smart and develop a strategy.

After matching patches, the students get to experience a simulated launch and re-entry sequence for SS2. After "dropping" from WK2, they hear the voice-over announce "3 seconds until rocket ignition ... 3, 2, 1". Then, the rockets fire and the students are "launched" into space. Thanks to the high-fidelity simulator pod (Figure 3.18), students really get the sensation that they're flying.

A few of my A4H colleagues have taken NASTAR's SSTP course and they all say it's a great program. But, do two days of training really qualify you to be a

Figure 3.18 Jose Miguel Hurtado sits in NASTAR's spaceship simulator. Courtesy: A4H

suborbital scientist? After all, there is no emergency egress training in NASTAR's program, so what happens if you're employed as a scientist on board a vehicle that requires you to wear a pressure suit in the event of an emergency? Well, there's an organization that's thinking along those lines and that organization is A4H.

Astronauts for Hire

A4H is one of the few organizations pioneering the effort to standardize astronaut training by developing a qualification standard for its members. It's a move that has upset some people in the industry but A4H's initiative isn't setting standards for the sake of setting standards. Nor, as some people think, is it trying to impose those standards on the industry; it's simply providing a safeguard for its members and value for money to its clients. After all, research institutions and other potential employers will require that their employees flying their payload possess the knowledge and skills necessary to perform the many tasks required of a scientist in the suborbital research environment. While vehicle-specific training will no doubt be delivered by suborbital launch providers, who will be responsible for providing the training for the scientists? Well, in the case of A4H, the scientists will be trained in accordance with the A4H Operations Specialist (OS) training program.

In developing the OS training program, the A4H Training Committee decided that each OS will be required to successfully complete specific training prior to flying their missions. Very broadly, it was decided that, at a minimum, each OS be trained to operate safely within the vehicle so they will know how to respond to both planned and anomalous events. Also, prior to each mission, it was decided that each OS should receive vehicle and mission-specific training to cover all phases of flight by practicing the operational and procedural simulations of each mission phase. Also, it was agreed that each OS be familiar with the flight characteristics of the vehicle so they would be proficient in nominal and non-nominal flight conditions (such as abort scenarios and emergency operations). They would also be required to have a competent understanding of generic vehicle systems, vehicle characteristics, and vehicle capabilities, as well as operational, malfunction, and contingency procedures. After much discussion, the A4H Training Committee decided on some core qualification requirements, the first of which was completion of the NASTAR SSTP (or equivalent), which is an occupation-specific course that provides high-fidelity training in all aspects of the suborbital environment.

In addition to the NASTAR course, prospective OSs are required to complete a zero-G (Figure 3.19) flight – an experience that provides a relaxed introduction to the microgravity environment and indoctrinates OSs to the challenges of performing tasks in a dynamic flight environment. Another check that's required is the PADI or NAUI Open Water Dive Course; the Training Committee agreed that scuba-diving

Figure 3.19 A4H member Laura Stiles performs science on board a zero-G flight. Courtesy: A4H

Figure 3.20 A4H members performing emergency egress training from Survival System's helo dunker. Courtesy: A4H/Survival Systems

Figure 3.21 The author preparing to perform unusual attitude training in the back seat of the Hawk. Courtesy: Author

Figure 3.22 A4H members Brian Shiro, Christopher Altman, and Jason Reimuller prepare for sea survival training. Courtesy: A4H/Survival Systems.

is an inexpensive analog for microgravity and serves as valuable training. To demonstrate their proficiency in emergency egress, OS candidates are required to complete basic egress training such as the helo dunker (Figure 3.20), which teaches them how to react in a timely manner while under stress (while being drowned). Another stressful training increment is aerobatic or unusual attitude training (Figure 3.21), which teaches the candidate how to adapt to rapidly changing G-loads. In between completing their training, A4H OSs work towards completing four core academic modules: Human Performance in Space, Life Support Systems, Spacecraft Systems Engineering, and the Space Environment. Once they've checked all the boxes, they're qualified OSs and eligible for flight assignment. For those who would like to beef up their résumé, A4H recommends courses such as accelerated freefall and survival training (Figure 3.22), which are intended to prepare candidates for orbital flights.

ORBITAL FLIGHTS

At the time of writing, the only means for the general public to experience orbital spaceflight is via Space Adventures, who have a joint commercial venture between the Russian Aviation and Space Agency (Rosaviakosmos) and Rocket Space Corporation Energia (RSC Energia). To date, there have been seven high-profile

72 Medical Qualification and Training

Figure 3.23 Robert T. Bigelow stands next to a model of Bigelow Aerospace's inflatable space station. Courtesy: Bigelow Aerospace

customers who have flown to the International Space Station (ISS) and, recently, Space Adventures announced plans for additional orbital flights and circumlunar flights for an estimated fee of $100 million. In the next few years, another orbital option may be available thanks to Bigelow Aerospace, which plans to build a private orbital station based on NASA's inflatable space habitat (Figure 3.23) technology.

ORBITAL MEDICAL RISKS

Those wealthy spaceflight participants flying to the ISS will have to abide by the medical standards developed by the five partner space agencies of the ISS medical program. One set of these standards addresses paying spaceflight participants; these standards permit flights up to 30 days in length and reflect the traditional philosophy behind government space-agency medical standards, which has been to ensure operational safety by excluding a large number of medical conditions. This is done simply to avoid any negative impact on mission success due to medical conditions, to limit the potential risk of medical events occurring during flight, and to enable long-term medical clearances in support of astronaut career longevity. However, this philosophy flies in the face of the commercial space industry, whose philosophy is to maximize the number of potential space passengers by *limiting* the number of disqualifying medical conditions. Obviously, some rethinking is required.

Since the beginning of manned space exploration, professional astronauts have been subjected to rigorous medical selection before being allowed to fly on orbit. From a medical fitness perspective, these super-fit astronauts obviously can't be considered as a representative sample of the general population, which is worrying for those hoping for a relaxing of the medical standards; that's because even some of these professional astronauts, who have been subject to some of the most thorough medical selection tests imaginable, have experienced ground and in-flight medical events. For example, a study of astronauts involved in 106 Space Shuttle missions (STS-1 through STS-108) between April 1981 and December 2001 [12] covering over 5,496 flight days indicated that 98.1% of men and 94.2% of women reported 2,207 separate medical events or symptoms during flight. Reported medical events or symptoms included everything from space adaptation syndrome (39.6%) to injuries and trauma (8.8%) and from infectious diseases (1.3%) to endocrine, nutritional, metabolic, and immunity disorders (0.1%). And don't forget the 14 fatalities as a result of the *Challenger* and *Columbia* accidents. The bottom line is that orbital spaceflight exposes individuals to an environment that is much more dangerous than suborbital flight, and here's why.

Acceleration

We're already familiar with the types of acceleration and the preferred direction of G-forces but we haven't discussed the medical conditions that could be adversely impacted by exposure to acceleration. Of course, some medical conditions may be aggravated by suborbital flight but, if there's a problem during suborbital flight, it's only a matter of minutes before you're back on the ground; once you're in orbit, your nearest trauma care center will be a long way away. A discussion of the number of medical conditions that could be seriously aggravated by acceleration is well beyond the scope of this book because there are just so many! First, there are various cardiovascular pathologies such as congenital heart diseases, valvular heart diseases, cardiomyopathies, pericarditis, myocarditis, endocarditis, ischemic heart diseases, dysrhythmias, and aortic aneurysm. Then there are cerebrovascular diseases such as stroke, transient ischemic attack (TIA), intracranial bleed, intracranial aneurysm, and cavernous angiomas [3]. Next are cerebral tumors, which, if compromised, could lead to loss of consciousness or recurrent syncope. Those with musculoskeletal disorders such as symptomatic cervical arthritis, recent spinal injury, severe osteoporosis, spondylolysis, spondylolisthesis, herniated nucleous pulposus, non-healed displaced fractures, or non-reduced dislocations of large joints probably won't be allowed to fly either. Nor will those with ophthalmologic disorders such as retinal detachment, hemorrhages, or other retinal vascular problems; individuals with high degrees of myopia (less than –6 diopters) are probably at risk as well because of the increased risk of retinal detachment [3]. Those prone to severe chronic dizziness, positional vertigo, motion sickness, or other vestibular/orientation problems will also be relegated to the no-fly list, as will those with recent intra-cranial, intra-thoracic, or intra-abdominal trauma. The list is by no means extensive and, remember, we're just talking about acceleration.

Barometric pressure

The immediate risks of rapid and explosive decompression also hold true for those flying on orbit but, on orbit, these risks may be compounded by certain medical conditions. For example, if a rapid decompression event occurred during a suborbital flight, the crew and passengers would be exposed to a decrease in barometric pressure for only a few minutes – an event that would be survivable even for those with medical conditions that could be adversely affected by hypoxia. On orbit, however, a rapid decompression event could mean a death sentence. Just like acceleration, there are a number of medical conditions that can be exacerbated by exposure to decreased barometric pressure. First, there's a myriad of ear, nose, and throat (ENT) disorders such as allergic rhinitis, severe otitis media, severe otitis externa, and recurrent bleeding from the nasopharynx, to name just a few. Then there are all the pulmonary disorders such as pulmonary vascular disease, pneumothorax, pulmonary fibrosis, active asthma, emphysema, and chronic obstructive pulmonary disease (COPD). Other conditions that could be affected include a shopping list of cardiovascular pathologies such as congenital heart diseases, valvular heart diseases, cardiomyopathies, pericarditis, myocarditis, endocarditis, and dysrhythmias [3].

Microgravity

One of the most noticeable effects of spending time in orbit is the shift of body fluids towards the head [6], which manifests itself in the typical facial puffiness seen in astronauts and the frequent reports of nasal congestion and headache. Sensing that it's overloaded with fluid, the body responds by getting rid of the "extra" fluid, which, in turn, leads to a rapid decrease in plasma volume of between 10 and 15% (in addition, thirst and fluid intake also decrease). While these adaptive changes don't cause a problem, and are actually beneficial, on return to Earth, they result in a condition space physiologists call *orthostatic intolerance*, which just means astronauts find it difficult to stay upright for a few hours after landing; it's also one of the reasons you see Russian cosmonauts being carried off in a stretcher after landing – NASA astronauts just crank up the pressure in the pressure suits to give the illusion that they're not affected (seeing astronauts being carried away on stretchers or staggering to stay upright would be bad PR!).

While orthostatic intolerance isn't a problem for those with healthy cardiovascular and neurovestibular systems, for those individuals with pre-existing heart disease and vestibular abnormalities, the condition may prove more challenging. And, since we're talking about the neurovestibular system, remember SMS? It was mentioned briefly in the suborbital section on "microgravity". During a suborbital flight, the microgravity portion will last only four or five minutes so, even if you get really, really sick, you'll have the comfort of knowing you'll be back on the ground soon enough. But, on orbit, this syndrome can persist for up to three days; that's an expensive three days if you've just paid $5 million for a week's stay. Then there's the return to Earth – an event that also proves confusing to the neurovestibular system; astronauts re-adapting to Earth's 1-G environment find it difficult turning corners, can barely walk in a straight line (it's one of the reasons some astronauts wear their NASA-issued launch diapers to bed for the first couple of days on return to Earth, in case they have a call of nature and can't make it to the bathroom), experience illusions of movement of self or surroundings, and are prone to suffering from nausea and vomiting.

As if SMS and discombobulation aren't bad enough, commercial astronauts working on orbit will have to contend with a decrease in muscle mass and strength along with neuromuscular changes including a decrease in the fatigue resistance of muscles and decreased neuromuscular efficiency. The changes occur because muscles no longer have to work against the force of gravity. Once again, this won't be a significant problem for a healthy passenger, but the potential impact on an individual with pre-existing neuromuscular problems? Well, nobody really knows.

Then there's the problem of back pain and bone loss. Astronauts often complain about back pain during space flight and then again on returning to Earth. In orbit, the pain is probably due to stretching of spinal ligaments and tendons with the unloading of gravity. This unloading causes another serious medical issue: bone loss. In fact, the rate of bone loss is approximately 1–2% mass loss per month in the lower lumbar region. Again, this isn't likely to be a problem for a healthy passenger, but it could be a problem for someone with osteoporosis.

Also implicated in bone loss is the renal system because, as calcium is reabsorbed from the bone, urinary calcium levels also increase, which results in an increase in urinary phosphate and uric acid [3]. More worrying for flight surgeons and their passengers, this increase in urinary calcium, phosphate, uric acid, and decrease in urine volume can lead to supersaturation of the urine and an increased risk for the formation of renal stones: more bad news!

Bone isn't the only thing orbital astronauts lose; a decrease in red blood cell (RBC) mass is also observed post flight. It's thought that this occurs because, as plasma volume decreases during flight, hemoglobin concentration increases, causing a decrease in erythropoietin. Destruction of recently formed RBCs then reduces the level to that appropriate for life on orbit. When the astronauts return to Earth, the body restores the plasma volume that was lost during flight, a decrease in hemoglobin concentration occurs, and an increase in erythropoietin is observed; the syndrome, like so many observed in spaceflight, has its own name: spaceflight anemia.

Microgravity also produces physiological changes that may influence the uptake and effectiveness of medications. For example, we know that blood and bodily fluids shift to the upper part of the body; as a result, gastric emptying, intestinal absorption, and renal excretion rates are reduced. How these changes interact to affect specific medications is unknown, but the effects would be especially relevant to those individuals who need to take medications regularly due to preexisting medical conditions.

We've seen that extended stays in orbit produce all sorts of adverse effects, even in healthy individuals, so it shouldn't be too surprising that there are several medical conditions that could be adversely impacted by exposure to microgravity. First there are the various cardiovascular pathologies such as congenital heart diseases, valvular heart diseases, cardiomyopathies, pericarditis, myocarditis, endocarditis, ischemic heart diseases, and dysrhythmias. Then there are the cerebrovascular diseases such as stroke, transient ischemic attack (TIA), and intracranial aneurysm; anyone with a history of recent intra-cranial surgery will also be excluded. Those with endocrine disorders such as diabetes insipidus, hypothyroidism, hyperparathyroidism, or diabetes mellitus will also be outside the inclusion criteria for orbital flight, as will those with a history of severe chronic dizziness, positional vertigo, motion sickness, or other vestibular problems. Because of the atrophying effects on the muscles and bones, those with musculoskeletal disorders such as progressive or severe atrophy of the muscles, myasthenia gravis, muscular dystrophy, or severe osteoporosis will also be excluded. And, due to the risks of renal stones, individuals with renal disorders such as a history of urinary calculus won't make the cut.

Ionizing radiation

Ionizing radiation has enough energy to cause genetic damage that can lead to cellular death or dangerous mutations. The longer you're in space, the more damage this radiation can do. And don't think you'll be protected by the walls of the spacecraft; high-energy particles can penetrate through thick solid matter, including the hull of a space vehicle. And, unlike the weather forecast on Earth, predicting when bursts of radiation may strike is patchy at best. Now, the degree of radiation exposure is dependent on altitude and inclination, with radiation exposure being greater the higher the inclination. Obviously, orbital operators will fly their space stations and spacecraft at low altitudes to reduce radiation exposure, but how much will an orbital passenger be exposed to? Well, back in the days of the Space Shuttle, the radiation dosage ranged from 0.6 to 1.5 mSv/day, so the only medical condition that might preclude someone from flying in orbit is pregnancy.

Noise and vibration

There was a very good reason why NASA kept people five kilometers away when the Shuttle launched. It's because the Shuttle generated 215 dB; 200 dB will kill you. Think I'm exaggerating? Way back in the Apollo era, NASA had a crew that went around the launch pad area after a Saturn V launch to pick up all the dead animals killed by the sound pressure. Along with the noise comes vibration; some Shuttle astronauts compared a launch to being inside a train wreck! Obviously, exposure to high magnitudes of continuous vibration has the potential to cause significant injury, which is why sending those with musculoskeletal disorders (symptomatic cervical arthritis, recent spinal injury, severe osteoporosis, and non-healed displaced fractures) into orbit probably isn't a good idea.

Temperature and humidity

The lack of an atmosphere in space exposes space vehicles to extremely cold and hot temperatures, which represent a potential hazard for the flight crew and passengers, who must rely on the proper operation of the cabin heating, air circulation, and cooling systems. If any one of these systems fails to perform optimally, then performance will be affected. Even a drop in the humidity can result in a dry atmosphere, leading to eye irritation and nosebleeds, while a malfunctioning heating system could lead to occupants suffering from heat or cold illnesses; heat illnesses include heat stroke, heat hyperpyrexia, heat syncope, heat exhaustion, heat cramps, and chronic heat fatigue, while cold illnesses include hypothermia, frostnip, and frostbite. Given the dire consequences of a recalcitrant heating and cooling, flight surgeons will screen out those with chronic hypotension or hypertension, sweat gland abnormalities, previous severe heat illnesses, and impaired sweating due to spinal cord injuries or autonomic neuropathy.

Behavioral

"Any sane astronaut will feel the fear, or concern just prior to liftoff. If they don't admit they are lying to you."

Greg Johnson, NASA astronaut

The number is bleak and every astronaut who flies into space knows it: there is a one in 75 chance they will not return home. At least that was the risk back in the days of the Shuttle and there is no reason to believe the odds will be much better over the next 10–15 years. Perhaps Greg Johnson, *Endeavour*'s pilot on STS-134, said it best when he compared launching on the Space Shuttle to going into combat.

In addition to the risks of launch, there are the psychological and interpersonal stressors related to isolation and confinement that can have a definite impact on individual behavior and performance. Then there's the stress of being in a confined space, the lack of privacy, and risk of injury or death. Given all these risks, it makes sense that those hoping to be launched into orbit be screened for behavioral abnormalities. So, those with psychiatric problems, such as personality disorders, adjustment and psychosomatic reactions, anxiety disorders, and phobias, will be screened out. Likewise, those with severe depression, acute psychotic reactions, bipolar disorder, nicotine addiction, or a history of recent alcohol and drug abuse will also be struck off the list.

Medical assessment of orbital crewmembers

By now, you should have an appreciation of just how fit and healthy you will need to be if you want to fly on orbit as a crewmember. So, what should the orbital medical assessment include? Well, to begin with, prospective orbital astronauts will be required to provide their medical history (including family history); this questionnaire will include plenty of check marks for those medical conditions that could be adversely impacted by exposure to the operational environmental of orbital spaceflight. After the medical history questionnaire, potential crewmembers will need to provide information about the use of implanted pacemakers and whether they've suffered from tuberculosis, hepatitis, AIDS, or other serious infectious disorders – all immediate disqualifiers. Next is a section about the use of medications, alcohol, and other drugs, followed by a self-report history of psychiatric and/or psychological problems. Females will have to provide information about the date of their last menstrual period, current pregnancy, and any recent spontaneous or voluntary termination of pregnancy. Once the self-report questionnaires have been completed, the potential crewmember can move on to the actual physical.

The flight surgeon conducting the physical will begin by checking vital signs such as heart rate, respiratory rate, temperature, and blood pressure. Then, they will examine the head, face, neck, and scalp, followed by a check on the nose, sinuses, mouth, throat, ears, and eyes. The lungs and chest, heart, vascular system, stomach, genitourinary system, upper and lower extremities, and spine will also be examined. These tests will be followed by a general neurological and general psychiatric evaluation. It's a long list, but it doesn't end there. Once the physical is completed, the prospective orbital crewmember will move on to the lab tests. These will include routine hematology, clinical chemistry, a urine analysis, resting electrocardiogram, chest X-ray, audiogram, and various pulmonary function tests.

While conducting the examination and reviewing the results of the tests, the flight surgeon will be looking for any abnormality that might disqualify a crewmember from flight (this is one of the reasons why astronauts have never liked flight surgeons). Disqualifying conditions might include obvious problems such as any significant deformities of the musculoskeletal system, diseases, illnesses, injuries, infections, tumors, or other physiological or pathological or psychiatric conditions that might result in an in-flight medical emergency, an in-flight death, or might compromise the health and safety of other crewmembers. The flight surgeon will also look for anything that might interfere with the donning, doffing, and operation of personal protective equipment, and any condition that might interfere with emergency egress procedures. Basically, any significant medical condition that might result in an unexpected in-flight medical emergency represents a potential risk to the safety of the flight.

Now, that's not to say that prospective crewmembers who have significant medical conditions will be black-listed from orbital spaceflight because special medical clearance can be given on a case-by-case basis; take Richard Garriott, for example. Garriott (Figure 3.24), who is the son of astronaut Owen Garriott (who flew on Skylab 3 and STS-9), is a British-American video-game designer and entrepreneur. He's also known by his alter egos Lord British in Ultima and General British in Tabula Rasa. On October 12th, 2008, Garriott launched aboard Soyuz TMA-13 to the ISS as a self-funded spaceflight participant and, in doing so, became the 483rd person to go into space.

However, to get into space, Garriott had to undergo corrective eye surgery, and fix his fused kidneys and liver hemangioma[7] to pass the medical tests. I met Garriott

---

[7] A liver (hepatic) hemangioma is the most common noncancerous tumor of the liver, which can occur at any time, but is most common in people in their 30s to 50s. Hemangiomas may cause bleeding or interfere with organ function, depending on their location. The tumor is usually not discovered until medical pictures are taken of the liver for some reason, such as a spaceflight medical!

Figure 3.24 Richard Garriott. Courtesy: Wikimedia

briefly in Los Angeles in May 2009, where he had given a presentation at the Aerospace Medical Conference. As he spoke, there seemed to be a strange disconnect between the tall, fit guy who was speaking and the medical procedures he was describing. After all, Garriott had already spent time in other hostile environments, including a trip to Antarctica, and had dived to the *Titanic* on board a Russian minisub. As I watched him lift up his shirt to show the audience an unsightly scar across his stomach, I couldn't help thinking: "If this guy had to go through all this to be green-flagged for space, what hope is there for the rest of the population?" But don't lose hope. After all, Professor Stephen Hawking, who suffers advanced amyotrophic lateral sclerosis with significant mobility impairment, was able to safely participate in a zero-G flight. Of course, to do this, he was accompanied by his own medical team, who were involved in providing in-flight medical support. However, Hawking's zero-G flight demonstrated it may be possible to allow selected individuals with severe disabilities to participate in short-duration orbital space flights, although such a flight would be a complex – and expensive (!) – operational medical scenario.

Post-flight medical

As we've already seen, orbital flight has a profound effect on the human body, which means that the medical poking and prodding will likely continue even after crewmembers return from orbit. Typical post-flight medical issues will probably include some discombobulation (postural instability) and deconditioning, so it makes sense that crewmembers undergo a post-flight medical, not only to resolve any medical issue, but also for the collection of medical data, which could be used to establish orbital crewmember medical databases. These databases could support the development of outcome-based medical information support systems that could improve health-risk management techniques applicable to medical evaluation and granting of waivers.

ORBITAL TRAINING

The only company planning on sending astronauts into orbit in the near future is Bigelow Aerospace, which hopes to send sovereign customers to its inflatable habitats in the next few years. Hardly surprising, then, that companies offering orbital astronaut training are few and far between. But that's not to say they don't exist. Take Atlas Aerospace, for example. Established in 1999, Atlas Aerospace was created to train spaceflight participants to accomplish commercial space missions on board the Soyuz space vehicles and on the Russian segment of the ISS.

Atlas Aerospace

Unsurprisingly for a company that's been in the astronaut training business for more than a decade, their program is well balanced and provides all the training necessary for a mission. The program consists of three courses, which are discussed briefly here.

The first course is General Space Training (GST), which covers the principles of astronautics, the main components of a manned space complex, how to conduct research and experiments in orbit, the skills needed to work in a spacesuit, generic flight and parachute training, survival training, and an introduction to biomedical issues. The GST phase takes between 10 and 26 weeks, depending on the astronaut specialization. On completion of the course, the candidate receives the qualification of "Astronaut" and is ready to move on to Specialized Space Training (SST).

The SST phase, as its designation suggests, is a more detailed phase of instruction dealing with flight operations, life-support training, simulator training, extravehicular activity instruction, and biomedical and psychological training. The phase lasts between 8 and 10 weeks. Upon completion of SST, the candidate astronaut is admitted to Pre-Flight Space Training (PFST). During PFST, astronaut candidates hone their flight operation skills in various simulators and spend a lot of time studying mission-specific flight documentation. They also receive instruction on the locations of space-borne equipment, consoles and instruments, become familiar with emergency scenarios, practice docking and undocking procedures, and become proficient in the skills required to operate propulsion, orientation, communication, air locking, refueling, energy consumption, and the crew life-support equipment. Typically, the PSFT lasts between 8 and 16 weeks, after which the astronaut candidate is certified for flight.

Astronauts for Hire orbital training program

If you've been adding up all the weeks, you'll have calculated that the Atlas training can take between 26 and 52 weeks. That's a long time for the commercial spaceflight industry. Surely, it must be possible to train astronauts in a shorter *and* more economically acceptable timeframe? As A4H's Training Director, one of my tasks was to develop such a schedule. I figured you could squeeze all the training into five weeks and still prepare someone for orbital spaceflight (Appendix III).

The commercial spaceflight industry is still maturing, so the medical and training certification suggested in this chapter will undoubtedly be revisited as circumstances and new knowledge require. It's possible orbital training can be completed in fewer than five weeks and it may be that suborbital training will take longer than the two or three days suggested by Virgin Galactic. How the medical and training standards evolve will be driven by the industry and its vehicles, which is where we turn our attention to in the next section.

REFERENCES

[1] Banks, R.D.; Brinkley, J.W.; Allnutt, R.; Harding, R.M. Human Response to Acceleration. In J.R. Davis, R. Johnson, J. Stepanek, and J.A. Fogarty (eds), *Fundamentals of Aerospace Medicine*, 4th edn. Lippincott Williams & Wilkins, New York (2008).

[2] Banks, R.D.; Grissett, J.D.; Saunders, P.L.; Mateczun, A.J. The Effect of Varying Time at −Gz on Subsequent +Gz Physiologic Tolerance (Push–Pull Effect). *Aviation Space Environmental Medicine*, **66**, 723–727 (1995).

[3] Antuñano, M.J.; Baisden, D.L.; Davis, J.; Hastings, J.D.; Jennings, R.; Jones, D.; Jordan, J.L.; Mohler, S.R.; Ruehle, C.; Salazar, G.J.; Silberman, W.S.; Scarpa, P.; Tilton, F.E.; Whinnery, J.E. Guidance for Medical Screening of Commercial Aerospace Passengers. Technical Report No. DOT-FAA-AM-06-1, Federal Aviation Administration, Office of Aerospace Medicine, Washington, DC (2006).

[4] Buckey, J.C., Jr; Gaffney, A.F.; Lane, L.D.; Levine, B.D.; Watenpaugh, D.E.; Wright, S.J.; Yancy, C.W., Jr; Meyer, D.M.; Blomqvist, C.G. Central Venous Pressure in Space. *Journal of Applied Physiology*, **81**, 19–25 (1996).

[5] Foldager, N.; Anderson, T.A.; Jessen, F.B.; Ellegaard, P.; Stadeager, C.; Videbaek, R.; Norsk, P. Central Venous Pressure in Humans during Microgravity. *Journal of Applied Physiology*, **81**, 408–412 (1996).

[6] Thornton, W.E.; Moore, T.P.; Pool, S.L. Fluid Shifts in Weightlessness. *Aviation, Space Environmental Medicine*, **58**, A86–A90 (1987).

[7] Buckey, J.C.; Neurovestibular Effects. In J.C. Buckey, Jr (ed.), *Space Physiology*. Oxford University Press, New York (2006).
[8] Ortega, H.J.; Harm, D.L. Space and Entry Motion Sickness. In M.R. Barratt and S.L. Pool (eds), *Principles of Clinical Medicine for Space Flight*, 1st edn. Springer, New York (2008).
[9] National Council on Radiation Protection. Radiation Protection Guidance for Activities in Low-Earth Orbit. NCRP Report No. 132, National Council on Radiation Protection and Measurements, Bethesda, MD (2000).
[10] Townsend, L.W.; Fry, R.J. Radiation Protection Guidance for Activities in Low-Earth Orbit. *Advances in Space Research*, **30**, 957–963 (2002).
[11] Shoenberger, R.W.; Harris, C.S. Human Performance as a Function of Changes in Acoustic Noise Levels. *Journal of Engineering Psychology*, AMRL-TR-65-165 (1974).
[12] Antuñano, M.; Hobe, S.; Gerzer, R. (eds). *Position Paper on Medical Safety and Liability Issues for Short-Duration Commercial Orbital Space Flights*. International Academy of Astronautics, Paris (2009).

Section II
The Industry

4

Meet the Industry

Commercial spaceflight is any flight above 100-km altitude conducted and funded by an entity other than a government. In the Space Age of the 1960s and 1970s, the government space agencies of the Soviet Union and US pioneered space technology, supported by collaboration with design bureaus in the USSR and private companies in the US. In 1975, the European Space Agency (ESA) was formed and it adopted the same model of space technology development. Then, along came the defense contractors, like Lockheed Martin and McDonnell Douglas, who began developing and operating space launch systems derived from government rockets and commercial satellites. More recently, entrepreneurs such as Elon Musk, Jeff Bezos, and Robert T. Bigelow have begun designing and deploying competitive space systems such as suborbital space-planes and launching lightweight orbital rockets (Figure 4.1).

The successes of Elon Musk's company SpaceX and Bigelow Aerospace represent nothing less than a seismic shift in space operations. After all, during the early years of spaceflight, only nation states had the resources to develop and fly spacecraft, which meant launching astronauts into orbit was the monopoly of a small group of national governments. The first phase of private space operation was the launch of commercial communications satellites, thanks to the US Communications Satellite Act of 1962, which opened the way to commercial consortia owning and operating their own satellites, although these were still launched on state-owned launch vehicles. The next phase came on March 26th, 1980, when the European Space Agency (ESA) created Arianespace. As the world's first commercial space transportation company, Arianespace produces, operates, and markets the Ariane (Figure 4.2) launcher family.

On the other side of the Atlantic, the policy of the US was that NASA be the public-sector provider of US launch capacity to the world market. Initially, NASA subsidized satellite launches with the intention of eventually pricing Shuttle service for the commercial market at long-run marginal cost. Then, on October 30th, 1984, President Ronald Reagan signed into law the Commercial Space Launch Act (CSLA), enabling an American industry of private operators of expendable launch systems. This act was followed by the Launch Services Purchase Act (LSPA), which

86 Meet the Industry

Figure 4.1 SpaceX's Falcon-9. Courtesy: SpaceX

Figure 4.2 Ariane. Courtesy: European Space Agency

was signed into law on November 5th, 1990, by President George H.W. Bush. The LSPA, in a complete reversal of earlier Space Shuttle monopoly, ordered NASA to purchase launch services for its primary payloads from commercial providers whenever such services were required in the course of its activities:

> "With the advent of the ISS, there will exist for the first time a strong, identifiable market for 'routine' transportation service to and from LEO, and that this will be only the first step in what will be a huge opportunity for truly commercial space enterprise. We believe that when we engage the engine of competition, these services will be provided in a more cost-effective fashion than when the government has to do it."
>
> NASA Administrator, Dr Mike Griffin, November 2005

Commercial human spaceflight took a big step on January 18th, 2006, when NASA announced an opportunity for commercial providers to demonstrate orbital transportation services. The Commercial Orbital Transportation Services (COTS) was a NASA program to coordinate the delivery of crew and cargo to the International Space Station (ISS) by private companies. Under the agreement, NASA signed agreements with Rocketplane Kistler (RpK) and Alliant Techsystems (ATK) in 2006, although it later terminated the agreement with RpK due to insufficient private funding. Then, in December 2008, NASA awarded another round of contracts for cargo delivery to the ISS to Orbital Sciences and SpaceX to utilize their COTS cargo vehicles. The idea behind COTS was simple; instead of flying payloads to the ISS on government-operated vehicles, NASA would spend $500 million (less than the cost of a single Shuttle flight) through 2010 to finance the demonstration of orbital transportation services by commercial providers. Unlike previous NASA projects, the spacecraft were intended to be owned and financed by the companies themselves and to be designed to serve US government agencies and commercial customers. Under the COTS rules and regulations, the private spaceflight vendors were competing for four specific service capabilities:

Capability level A: External unpressurized cargo delivery
Capability level B: Internal pressurized cargo delivery
Capability level C: Internal pressurized cargo delivery, return, and recovery
Capability level D: Crew transportation.

One of the reasons for COTS was funding; in 2006, NASA was committed to realizing the Vision for Space Exploration (VSE), a program dedicated to returning astronauts to the Moon and venturing beyond the Moon to Mars. Unfortunately, NASA didn't have the money to achieve all the VSE's objectives – a situation that led then NASA Administrator Michael D. Griffin to say that NASA needed affordable COTS if they were to realize the goal of returning astronauts to the Moon. Then, when President Obama took office, the goalposts changed again, with VSE being abandoned and the Shuttle forced into retirement. COTS took on new significance because, with the Shuttle retired, the US found itself without an independent manned space capability.

Figure 4.3 SpaceX's Dragon capsule. Courtesy: SpaceX

The awarding of contracts to SpaceX and Orbital Sciences Corporation (OSC) in December 2008 called for a minimum of 12 missions for SpaceX and eight missions for OSC. Of the two companies, SpaceX quickly became the frontrunner, launching their first Falcon-9 rocket and a mockup Dragon capsule (Figure 4.3) on June 4th, 2010. This success was followed by the first resupply demonstration flight, Dragon C1, which took place on December 8th, 2010. This flight demonstrated the Dragon capsule's multiple orbit capability, ability to receive and respond to ground commands, and ability to gain and maintain directional alignment with NASA's Tracking and Data Relay Satellite System (TDRSS) communication system, which is used in conjunction with all manned spaceflight to the ISS.

In April 2011, NASA announced Round 2 of the CCDev awards. The latest beneficiaries of NASA's largesse were SpaceX ($75 million), Blue Origin ($22 million), Sierra Nevada Corporation ($80 million), and aerospace behemoth, Boeing, which landed a deal worth $92.3 million to refine the design of its CST-100 (Figure 4.4) crew capsule.

While the CCDev-2 awards underlined the strongly corporate nature of the program, they also emphasized the ruthless bidding process; of the 22 CCDev-2 bids NASA received, 18 weren't selected for second-round awards, including United Launch Alliance (ULA), Excalibur Almaz (EA), OSC, and United Space Alliance

Figure 4.4 Boeing's CST-100. Courtesy: Boeing

(USA). ULA was perhaps the biggest surprise of the CCDev-2 announcement because the company had been one of five first-round CCDev awardees and its launch vehicles factored significantly into the plans of other commercial crew development companies. EA, which has plans to use Russian Almaz spacecraft for commercial space flights, was a surprise finalist for CCDev-2; while there were few details about what EA was proposing for CCDev-2, it's likely the company will continue its commercial activities, although at what externally appears to be a slow pace. OSC, meanwhile, had made a big splash in 2010 with its commercial crew development plans, using a lifting body concept called Prometheus launched on an EELV, building upon interest in a commercial crew vehicle that dates back to the 1990s. Failure to secure a CCDev-2 award will put the company in a tough spot, although they will probably continue to stay in contention for future commercial

Figure 4.5 Bigelow Aerospace's space station. Courtesy: Bigelow Aerospace

crew awards. USA, the Boeing–Lockheed Martin joint venture that operated the Space Shuttle, had put forward a proposal to continue flying two of the orbiters, *Atlantis* and *Endeavour*, commercially. Not surprisingly, USA was not among the eight companies shortlisted for CCDev-2, and even company officials admitted the proposal was "an extremely long shot".

While government funding has traditionally been a strong stimulus for aerospace programs, from classified "skunk works" tech to pre-launch loan funding, there are those who wonder what may happen if the third round of CCDev is cancelled. After all, some of the Round 2 winners are reliant on government funding – even Boeing, which won the lion's share of the money. Here's what John Elbon, Boeing's CST-100 program manager, had to say when asked about what might happen if Boeing wasn't selected for the next round of funding:

> "If for some reason we weren't selected to continue in the next phase then it would be a very difficult decision for us to continue on our own, and most likely we would not."

Remember, this is Boeing! Not surprisingly, such uncertainty about budget tightening and looming political battles (NASA funding is always at stake in future budget negotiations as US presidential elections approach) forced some outside NASA to examine just how stable the agency is as an anchor customer and plan for the worst. Already, interest has come from countries seeking a way to launch their own missions without needing to develop the resources.

Bigelow Aerospace, which is developing a space station (Figure 4.5) of sorts, has already signed agreements with organizations in six countries: Japan, Singapore, UK, Sweden, Australia, and the Netherlands. Bigelow has also signed an agreement with Boeing – Bigelow will provide the destination and Boeing will provide transport in the form of its CST-100 capsule.

CCDev money gives the recipients a head start against non-funded rivals and, while few companies have given up completely after failing to win a CCDev award, it would seem that the near future probably belongs to the winners. However, this may not be the case because the CCDev-2 awardees are involved in the development of vehicles designed to ferry astronauts into low-Earth orbit (LEO) – an undertaking requiring long lead times. In contrast, many of the companies denied funding are suborbital companies, some of whom are close to realizing operational status.

SUBORBITAL OPERATORS

Before 2004, no privately operated manned spaceflight had ever occurred. The only private individuals to journey to space went as spaceflight participants on Russian Soyuz flights to Mir or the ISS. The Ansari X PRIZE was intended to change that by stimulating private investment in the development of spaceflight technologies. The June 21st, 2004, test flight of SpaceShipOne (SS1), a contender for the X PRIZE, was the first human spaceflight in a privately developed and operated vehicle. Hot on the heels of SS1, Richard Branson, owner of Virgin, and Burt Rutan, SS1's designer, announced that Virgin Galactic had licensed the vehicle's technology and were planning commercial space flights. A fleet of five craft (SpaceShipTwo, launched from the WhiteKnightTwo carrier airplane) were to be constructed and flights would be offered at $200,000 each.

Virgin Galactic

Virgin Galactic is a company within Richard Branson's Virgin Group, which plans to provide suborbital spaceflights, suborbital space science missions, and, eventually, orbital flights. The six-passenger spacecraft, SpaceShipTwo (SS2), described in detail in the following chapter, will be carried to about 16 km by a carrier aircraft, WhiteKnightTwo (WK2), at which point, SS2 will separate and continue to over 100 km. The time from lift-off of the White Knight booster carrying SS2 until the touchdown of SS2 after the suborbital flight will be about 3.5 hr. The weightlessness will last approximately five minutes.

Branson intends to run the first flights out of New Mexico before extending operations around the globe. To date, more than 400 individuals have signed up for

Figure 4.6 Artist's rendering of Spaceport America. Courtesy: Spaceport America

a flight, each paying $200,000. The New Mexico authorities are investing almost $200 million in Spaceport America (Figure 4.6), which features a 3,000-m runway and a space-age terminal and hangar building designed by Foster and Partners. In addition to Spaceport America, Virgin plans to operate out of Spaceport Sweden and possibly from a UK base: RAF Lossiemouth. Charged with building the fleet of Virgin Galactic spacecraft is The Spaceship Company (TSC), a new aerospace production company, founded by Sir Richard Branson's Virgin Group and Scaled Composites. TSC's initial launch customer is Virgin Galactic, which has contracted to purchase five SS2s and two WK2s.

While the business case for public space transportation has yet to be proven, ticket sales for suborbital flights at $200,000 a seat have been tallying up. According to Alex Tai, vice president of operations for Virgin Galactic, the business plan is for 50,000 people to visit space over a 10-year time period. Having worked on the Virgin Galactic project from its conception, Tai plans to fly the first commercial flight of the firm's spaceship as one of the pilots. Among current tasks, he's supervising the design and construction of SS2.

There are some in the business world that are still skeptical about commercial space travel – more stunt than a stable business, they say. However, Branson is keen on developing a legacy and becoming the first-ever spaceliner operator and owner, and, while the price of those seats is out of reach for most people, incremental improvements to the technology will result in a drop in the cost of access. How long that will take is difficult to say because it's a reasonable assumption that Branson will want to make money on his space venture and, to do that, he has to recoup some of his not inconsiderable investment; by the time the first systems are delivered to Virgin Galactic, something in the range of $150 million will have been spent, and that figure doesn't include the money spent on the operational structure, which includes facilities to handle early operations in Mojave, California, and at Spaceport America in New Mexico. Add all those costs together and you get a figure approaching $250 million to reach Virgin Galactic's operating point.

But, if the surveys forecasting the volume of suborbital space passengers are correct, then hundreds of spaceships may be needed. The revenue from these flights can then be leveraged to achieve cheap access to LEO, which is where the real market is. With that technology in hand, Virgin Galactic will set its sights on orbital destinations, space hotel stopovers, the Moon, and beyond, but, for now, the company has its sights set on suborbital flight. It's a story that has been followed by the media to such a degree that the public might be forgiven for thinking that Virgin Galactic is the only company in the space tourism game but, of course, that's not the case.

XCOR Aerospace

XCOR Aerospace is a private rocket engine and spaceflight development company based, like so many other commercial spaceflight ventures, at the Mojave Spaceport in Mojave, California. Headed by CEO, Jeff Greason, XCOR was formed by former

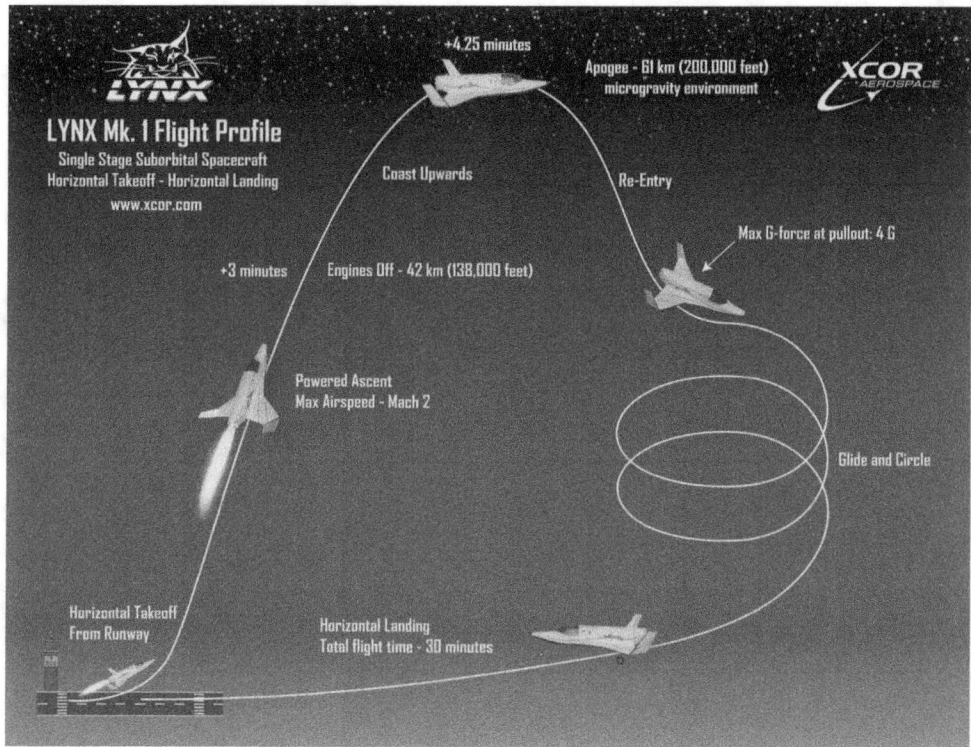

Figure 4.7 Flight profile of XCOR's Lynx spacecraft. Courtesy: XCOR

members of the Rotary Rocket engine development team in September 1999; the company evolved from working out of the chief engineer's tiny hangar to a team of 20+ employees housed in a 10,375-square-foot hangar. XCOR's star attraction is the Lynx, a stubby-winged vehicle capable of carrying a pilot and a passenger or payload on suborbital flights (Figure 4.7). Up to 50 test flights of Lynx are planned, along with numerous static engine firings on the ground. Once operational, it will be capable of flying several times a day, making use of reusable, non-toxic engines to help keep the space plane's operating costs low.

XCOR has flown a piloted rocket operations demonstrator aircraft more than 60 times without incident – flights that have proved the company's reusable rocket engines are capable of multiple flights per day and in-flight restarts. With nearly 4,000 engine firings and 500 minutes of run time on their engines, XCOR is at the forefront as a leader in the commercial space transportation industry, not only by virtue of its engineering prowess, but also as a result of their involvement in the passage of the Commercial Space Launch Amendments Act (CSLA) of 2004 (Jeff Greason served on the US Human Space Flight Plans Committee).

Unlike Virgin Galactic, which has the support of Branson's big bucks, XCOR's research and development funding has come mostly from precisely targeted commercial development programs, government research contracts, investment

96 Meet the Industry

Figure 4.8 Sierra Nevada's Dream Chaser. Courtesy: Sierra Nevada

capital from angel investors, and consulting services. These programs and contracts have led the company to the verge of being able to provide affordable suborbital launch services to academic, scientific, engineering, and observation-related markets.

SpaceDev

SpaceDev is a wholly owned subsidiary of the Sierra Nevada Corporation (SNC). SpaceDev's vehicle, the Dream Chaser, is a planned crewed suborbital and orbital vertical-take-off, horizontal-landing (VTHL) lifting-body spaceplane (Figure 4.8). The vehicle is planned to carry seven people to and from LEO, launching vertically on an Atlas V and landing horizontally on a conventional runway.

In common with SS2, Dream Chaser has a long history. The vehicle was publicly announced on September 20th, 2004, as a candidate for NASA's VSE and later for the agency's COTS. When the spacecraft wasn't selected under Phase 1 of the COTS Program, SpaceDev founder Jim Benson stepped down as chairman of SpaceDev and started Benson Space Company to pursue the development of the Dream Chaser. Then, in April 2007, SpaceDev announced it had partnered with the ULA to investigate utilizing the Atlas V booster rocket as the Dream Chaser's launch vehicle. This partnership was followed by another agreement when, on August 18th, 2008, SpaceDev signed a multi-year contract with Scaled Composites to assist Scaled in the development of a production rocket motor for SS2. Under the contract, SpaceDev has served as the lead rocket motor team member for SS2 and has collaborated with Scaled's internal design team to develop a production-ready hybrid rocket motor.

Also, SpaceDev has conducted ground tests on those motor components and has worked to assist Scaled in the full-scale rocket test program on the ground and during SS2 flight tests. Four months after the Scaled deal, SpaceDev was acquired by SNC, and the company began preparing its application for NASA's CCDev-1. On February 1st, 2010, the corporation was awarded $20 million in seed money under NASA's CCDev-1 for the development of the Dream Chaser (of the $50 million, Dream Chaser's award represented the largest share). This was followed by another $80 million the following year under CCDev-2.

SpaceDev's lion's share of the CCDev-2 cash wasn't surprising given the company's impressive array of technology partners, which enabled it to submit a strong proposal. Among the company's partners are Boeing Phantom Works, which will be constructing the test articles, the Charles Stark Draper Laboratory (guidance, navigation, and control), Aerojet (reaction control system technology), the University of Colorado (human-rating), and MDA (systems engineering). Thanks to these partners, progress will be quick.

Blue Origin

Blue Origin is a privately funded aerospace company set up by Amazon.com founder Jeff Bezos. The company was awarded $3.7 million in funding in 2009 under the CCDev program to develop suborbital concepts and technologies. Of particular interest to NASA was Blue Origin's innovative "pusher" Launch Abort System (LAS), a different type of abort system than the traditional tractor variety that pulls the crew vehicle to safety in case of an emergency. Focused primarily on suborbital spaceflight, the company has built and flown a test bed of its New Shepard spacecraft (Figure 4.9) design at their Culberson County, Texas, facility. The craft will launch vertically, like the classic rocket ship of science-fiction movies, and will land vertically as well. A notoriously secretive organization, Blue Origin's most recently publicized timetable states that New Shepard will fly manned in 2012. When operational, the vehicle will carry seven passengers into space several times a week.

In common with Virgin Galactic, Blue Origin was created by a billionaire: Jeff Bezos, of Amazon fame. As a teenager, Bezos won a trip to NASA's Marshall Space Flight Center by writing a paper on the effect of zero gravity on houseflies. Then, in 1982, he told the *Miami Herald* that he hoped to one day put space hotels in orbit. Not surprising, then, that Blue Origin's vision doesn't stop at the suborbital and ballistic trajectory stage; the company has a 20-year plan that envisions everything from hotels to space colonies and trips to Mars. With a net worth estimated at $1.7 billion, Bezos may avoid the fiscal troubles that doomed many of his predecessors with similarly lofty dreams.

Bezos's secretive Blue Origin operation is headquartered in a warehouse on East Marginal Way in Seattle. The company has no listed phone number, and Bezos and his spokespeople steadfastly decline to reveal many details. One detail the company couldn't keep under wraps was Bezos's purchase of several hundred thousand acres near Van Horn, Texas, for the purpose of building a spaceport. Van Horn, a one-

98 Meet the Industry

Figure 4.9 Blue Origin's New Shepard spacecraft. Courtesy: Blue Origin

horse town of 3,000 hardy individuals, is a dusty blip on a Texas highway, memorable only for the relief if offers after driving interminable miles of empty road; the town's selection as a spaceport has to be one of the unlikeliest twists in space exploration. Of course, for a "skunk work's"-type operation like Blue Origin, the remote location is perfect, although, given what happened in August 2011, Bezos may have wished the site had been even more secluded; on August 24th, 2011, an unmanned Blue Origin spacecraft was performing a developmental test flight from the company's West Texas spaceport when ground personnel lost contact and control of the vehicle. While travelling at Mach 1.2 at an altitude of 45,000 ft, a flight instability drove an angle of attack that triggered Blue Origin's range safety system to terminate thrust on the vehicle. It wasn't the outcome the Blue Origin team wanted, but, at the time of writing, they were already working on their next development vehicle.

ORBITAL OPERATORS

Orbital space tourism is a niche industry at the moment; to date, only seven people have paid to launch into Earth orbit, and they've reportedly paid between $20 million and $35 million for the experience. Those are not the numbers of a thriving industry. But things could change dramatically if prices drop to about $500,000 per seat or so. Today, the only agency selling orbital trips is the Virginia-based company, Space Adventures, which sells seats aboard Russian Soyuz vehicles. It's not surprising that Space Adventures is the only game in town, because the pool of potential customers willing and able to pay $30 million for a space jaunt is pretty small. However, a landmark Futron/Zogby space tourism study found, among other things, that 30% of the polled millionaires would be willing to spend $1 million for a two-week orbital trip. That equates to about 9,000 people; good numbers, but not good enough to support a thriving industry. But, according to Futron, if ticket prices can be reduced to $500,000, the global customer base is 14,000 in the conservative case, but nearly 225,000 in the optimistic scenario, which means there would be a requirement for more than just suborbital rockets.

As we've already seen, NASA has doled out millions of dollars to companies to develop orbital transportation methods so they don't have to continue paying the Russians $58 million a seat to ferry their astronauts to the ISS. However, these companies, while interested in being a customer of NASA, also have private spaceflight as part of their business plan. Space Adventures, for its part, has already signed a contract with Boeing to fly people to orbital space on its CST-100 vehicle, and those trips could start as soon as 2015. In fact, the competition between the NASA-funded private spacecraft companies point to a new race to orbital space that could help drive costs down substantially, through competition and the development of new, more efficient technologies. Only time will tell whether prices drop and by how much and how fast, but the signs point encouragingly towards the dawn of a new era in human spaceflight; thanks to NASA, these private companies could open the heavens to more and more folks who don't have $30 million lying around:

> "This really is what I would call an inflection point for human spaceflight. This change in human spaceflight will not only regain leadership for the United States, it will take us light-years ahead of everybody else."
> Bretton Alexander, president of the Commercial Spaceflight Federation

Bigelow Aerospace

Las Cruces is a Western frontier settlement located on the edge of the Chihuahuan Desert, along the banks of the Rio Grande. The history of the town, which is about 380 km south of Albuquerque, is neatly displayed at the city's Farm and Ranch Heritage Museum, which presents examples of folk art, farming tools, and wagons once drawn by horses. For the past six years, the museum has also served as the site of the International Symposium for Personal and Commercial Spaceflight (ISPCS) – an event that concerns a very different type of transportation and a frontier hundreds of kilometers above the museum. The 2010 symposium was the biggest yet, attended by 400 industry professionals – evidence of the growing interest in the commercial spaceflight business. For the first time at ISPCS, Bigelow Aerospace had a marketing booth, displaying models of its space habitats and information about its ambitions beyond Earth orbit.

It takes a certain boldness to pioneer a new market, especially in a capital-

Figure 4.10 Bigelow Aerospace's BA-330 module. Courtesy: Bigelow Aerospace

intensive, highly regulated industry like commercial spaceflight, but that's what Bigelow Aerospace is doing. Owned by Robert T. Bigelow, the billionaire founder of the Budget Suites of America hotel chain and would-be space real-estate mogul, the Las Vegas-based company has developed and launched two subscale prototypes – Genesis 1 and 2 – and has plans to launch larger modules in the near future to support applications ranging from commercial research to "rent-a-stations" for governments. On display at ISPCS was an exhibit that demonstrated just how large the company's ambitions really are: "Space Complex Alpha", a baseline space station, which uses one BA-330 module (Figure 4.10) and two smaller modules, linked together with a docking node and a propulsion bus. The company also showed off an even larger concept – Space Complex Bravo – using four BA-330 modules with two docking nodes and two propulsion buses.

However, even the voluminous Bravo was dwarfed by another concept on display: the BA2100. This module, as its name suggests, will have 2,100 cubic meters of habitable volume, over six times that of the BA-330. The cutaway model showed four decks arranged lengthwise inside the module, providing ample room for crew quarters and lab space. In keeping with its secretive nature (although not as obsessively clandestine as Blue Origin), Bigelow disclosed few other details about the concept module beyond the fact it would be too large to be launched on an existing rocket.

Despite these bold plans, the company is still relatively modest, with just over 100 staff on the payroll (SpaceX, in contrast, has about 1,000 employees). In the near term, the company is focused on building the Sundancer and BA-330 modules that will form the basis of its orbital habitats. While the company has often been portrayed as a developer of "space hotels", Bigelow maintains their primary markets are research and "sovereign clients". Just prior to the ISPCS, the company revealed it had signed memoranda of understanding with six countries: Australia, Japan, the Netherlands, Singapore, Sweden, and the UK. Bigelow believes as many as 50 countries will be interested in using the habitats, although, because of the recession, making deals has been more challenging.

A bigger hurdle for Bigelow Aerospace is finding the means to access the habitats once launched; once a BA-330 habitat is assembled, it will need a steady stream of spacecraft launched to it, carrying crews and cargo needed to maintain and utilize the facility. That's a big obstacle for the company because there are no commercial options for LEO except Russia's Soyuz and Progress spacecraft. Because of the dearth in commercial orbital transportation systems, Bigelow is a supporter of NASA's efforts to develop commercial crew transportation systems designed to service the ISS, but which could also be used to access Bigelow facilities. At ISPCS, Bigelow revealed that he had partnered with Boeing to develop the CST-100 crew spacecraft, which Boeing has been working on thanks to CCDev funding. He also indicated the company was very interested in SpaceX's development of its Dragon spacecraft. For their part, Boeing and SpaceX are interested in Bigelow Aerospace because the company offers the promise of a sustained, large market for space transportation services; once Alpha station is launched, it will need six flights a year and, with the launch of a second, larger station, that number would grow to 24. However, the company's vision extends beyond LEO; Bigelow's big plans for human spaceflight extend to deep space, including lunar outposts. Asked when he hoped to have his first space station operational, Bigelow said it was a fair bet they would be servicing clients by 2015, although he acknowledged this date could be impacted by the availability of a transportation system. But, with SpaceX's recent test launch of its Dragon capsule, Bigelow may not be waiting very long. And, once that new phase of space exploration and development is kicked off, the company will be well on its way to developing the new high frontier.

SpaceX

Space Exploration Technologies Corp. (SpaceX) is a commercial space transport company founded by PayPal co-founder, Elon Musk, who has invested more than $100 million of his own money. Originally based in El Segundo, SpaceX now operates out of Hawthorne, California, where it has nearly doubled in size every year since it was founded in 2002: from 160 employees in 2005 to more than 1,100 in 2010. As well as developing the Falcon-1 and Falcon-9 rockets, the company is also developing the Dragon (Figure 4.11) spacecraft to be carried to orbit by the Falcon-9 launch vehicle. In December 2010, SpaceX became the first private company to successfully launch, orbit, and recover a spacecraft (a Dragon).

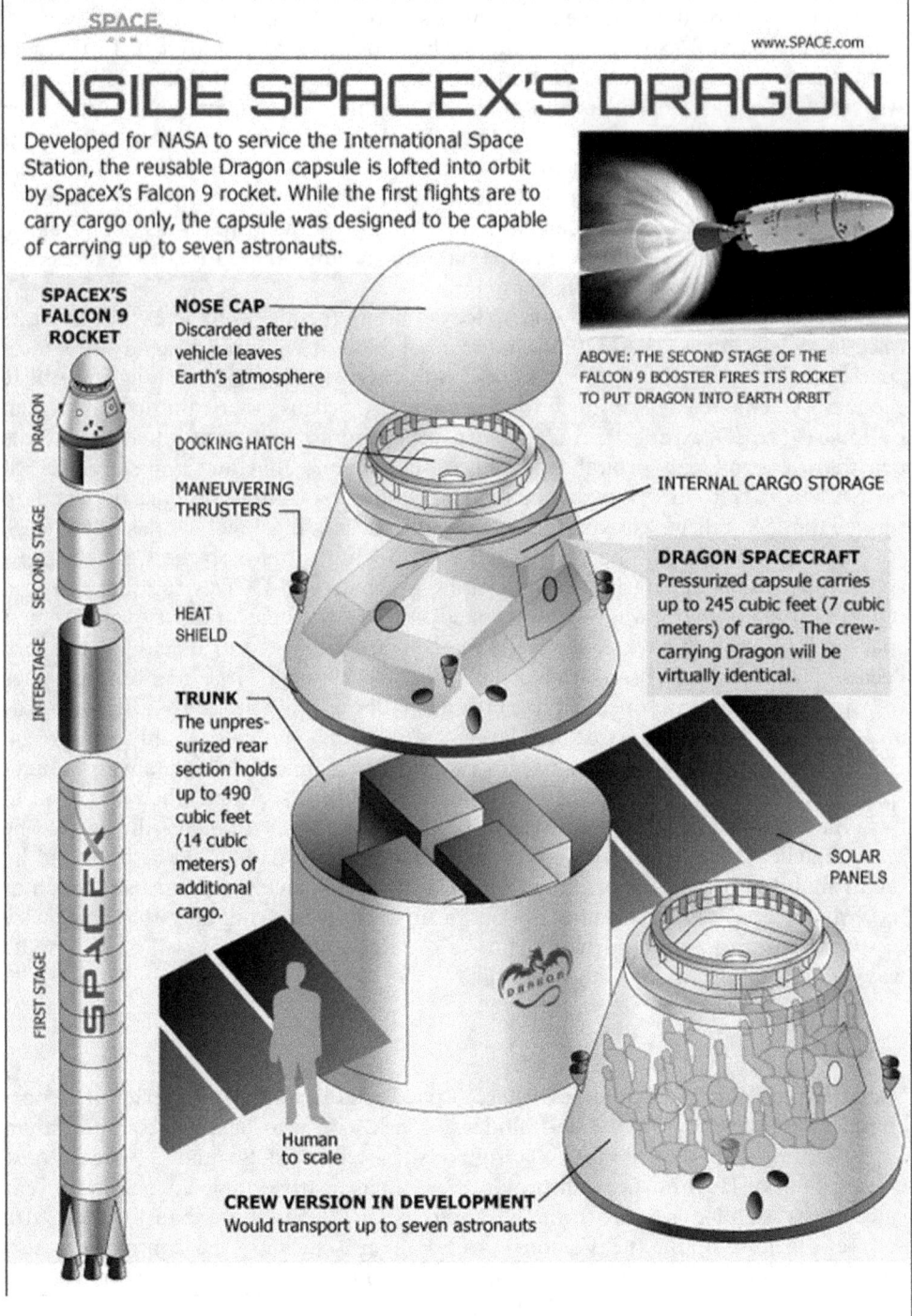

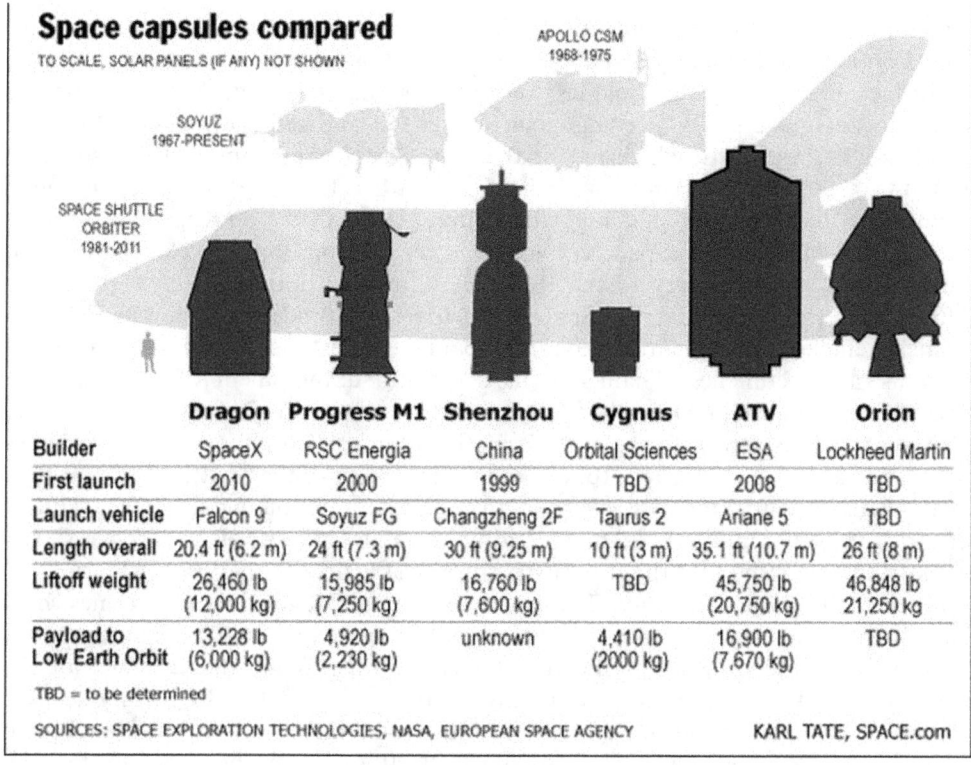

Figure 4.11 Infographic of SpaceX's Dragon capsule. Courtesy: Space.com

For Musk,[1] SpaceX was a means to seek a new direction in life, after having bought the idea for PayPal and sold it to eBay for $1.5 billion. Thinking philanthropy too staid, and being too young to retire (he's only 40), Musk picked cheap and reliable access to space as an ambition worthy of his talents. Ask any space policy analyst and they will tell you that launching rockets is an enormously difficult business to make money in. Those same analysts will also tell you that new rocket science has a high mortality rate, but Musk seems to have avoided the pitfalls. To realize his goal to make a business out of inexpensively launching satellites and humans into orbit, Musk has built the space equivalent of an all-American Volkswagen, a dirt-cheap rocket that launches into space, carrying payloads at a

---

[1] A native of Pretoria, South Africa, at age 17, Musk moved to Canada, where he briefly attended Queens University in Kingston, Ontario. He later transferred to the University of Pennsylvania, where he received two undergraduate degrees: one in theoretical physics and the other in business. He enrolled in the graduate physics program at Stanford in 1995 but dropped out to become an internet entrepreneur. His first big hit was Zip2, a Web-based ad company that he sold in 1999 for $307 million. He moved on to another Web venture, involving electronic payments over the internet, a business that became PayPal.

fraction of the cost of his competitors. In common with Bigelow, Musk has set his sights on making humanity a space-faring civilization.

In fact, it is exactly this boldness of ambition that sets SpaceX apart from the other rocket makers. To meet his goal of a cheap and reliable rocket, Musk has produced basic designs, which means there are fewer opportunities for systems to fail. It's a strategy that optimizes simplicity rather than performance; in short, economy is everything. It's a strategy that impressed NASA – on August 18th, 2006, the agency announced SpaceX had won a $278 million COTS contract to demonstrate cargo delivery to the ISS with a possible option for crew transport. Under the terms of the contract, SpaceX was to use the funds to develop its Falcon-9 launch vehicle, with incentive payments paid at milestones culminating in three demonstration launches. Thanks to its successful development of the Falcon-9, SpaceX announced in December 2008 that it had won a Commercial Resupply Services (CRS) contract, for at least 12 missions for US$1.6 billion to carry supplies and cargo to and from the ISS, following the retirement of the Shuttle. Then, in June 2010, Musk's company was awarded the largest ever commercial space launch contract (US$492 million) to launch Iridium satellites using Falcon-9 rockets. The same month saw the first flight of a structural test article of the Dragon during the maiden flight of the Falcon-9. Though the mockup Dragon lacked avionics, heat shield, and other key elements, an operational Dragon spacecraft was launched six months later and returned to Earth after two orbits. While the successes of the Falcon-9 and the Dragon may eventually assure US access to LEO, Musk's sights are set on a more distant goal: Mars. Landing astronauts on the Red Planet is an objective Musk aims to achieve in the next 10–20 years, and he hopes to do it by using the brawny lifting power of a 300-ft-high behemoth – dubbed the Falcon-XX – capable of hoisting 155 tons in LEO. To turbo-boost the astronauts to Mars, Musk plans to go back to the future and resurrect the NERVA (Nuclear Engine for Rocket Vehicle Application) – there are many who think that, if the Nixon administration hadn't killed the NERVA R&D effort in 1970, there would be astronauts walking on Mars today.

Excalibur Almaz

Another space exploration company with plans to open up a new era of private orbital space flight is Excalibur Almaz (EA). EA is an initiative that joins Russian space technology expertise with an international private enterprise to create a commercial offering of orbital spaceflight services. The company plans to offer week-long orbital space flights beginning as early as 2013, which is a big leap beyond the suborbital flight market targeted by most other private space companies. EA plans to tailor space missions to accommodate customer objectives including exploration, cargo transportation, and experimentation. On some missions, spacecraft and space stations will provide platforms for microgravity scientific experiments, potentially serving the needs of research and academic institutions. It sounds ambitious and you may be wondering how a company can design and develop so much infrastructure. Well, EA takes advantage of surplus

Soviet spacecraft. You see, EA's CEO, Art Dula, who was among the first Westerners to gain access to Russian space facilities, saw the wisdom to see the value in leftover Soviet spacecraft. With over 30 years' experience as an attorney specializing in aerospace, export control, and intellectual property law, Dula has served as a NASA consultant on Space Shuttle contracts and as a legal advisor to the US Congress. He also served as a Director and General Counsel for several aerospace companies, including Eagle Aerospace, Inc., which contracted to NASA and other aerospace companies. Dula's impressive credentials are supplemented by EA's impressive board of advisors, including former Johnson Space Center Director, George Abbey, former Kennedy Space Center Director, Jay F. Honeycutt, former Space Shuttle astronaut and VASIMR plasma rocket inventor, Franklin Chang-Diaz, former French astronaut Jean-Loup Chrétien, and former cosmonauts, Vladimir Titov and Yuri Glazkov.

Using Dula's business acumen, EA is employing elements of the legacy Russian space program known as "Almaz".[2] Working in technical cooperation with the Almaz program creator and many of the engineers and managers who oversaw the program from its beginning, EA is modernizing key elements of the Almaz space system with assistance from leading international aerospace contractors. Because the key elements of the original Almaz system have already undergone validation and flight testing, the development, design, testing, and engineering timeline is significantly streamlined, providing enormous cost advantages to customers awaiting the opportunity to fly in space.

Unlike many commercial space companies that crowd the Mojave Desert, EA, established in 2005, is incorporated, headquartered, and registered on the Isle of Man in the British Isles, although it has offices in Houston and Moscow. The reusable re-entry vehicle (RRV) – the TKS – vaguely resembles a cross between the American Gemini and Apollo capsules. They're equipped to carry three passengers or operate autonomously but, unlike the American capsules, the Almaz capsules (Figure 4.12) are reusable up to 100 times and can be launched atop any of several rockets. In common with the Soyuz, they use parachutes and retrorockets to return to Earth, and have soft-landing engines that fire just prior to touching down on land, although water landings are also possible.

---

[2] Almaz was a military space station project conceived in 1964 by NPO Mashinostroyeniya (Machine Building Scientific Production Association) in Moscow. A 20-ton "fortress in space" (presumably weaponized) spacecraft was to have operated for a couple of years to provide photographic and radar reconnaissance images. A special supply transport spacecraft, the TKS, would arrive at the spacecraft with cargo and cosmonauts. Its crew capsules were intended to be reusable up to 10 times. The Almaz program was approved in 1967. The first Almaz that was launched was called Salyut-2, launched into orbit at 0900 UT on April 3rd, 1973.

Figure 4.12 Excalibur Almaz capsule. Courtesy: Excalibur Almaz

SPACEPORTS

At the risk of stating the obvious, spaceports are to the commercial-spaceflight industry what airports are to the aviation industry. Spaceports also happen to be expensive to build, so the number of spaceports and their capabilities serve as a useful index of the health of the industry.

While many will be familiar with Spaceport America,[3] the New Mexico-based hub

---

[3] Visitors to New Mexico can already view Spaceport America up close because the world's first commercial spaceport opened its doors in May 2011. The spaceport, designed by renowned British architects Fosters and Partners, is situated in the New Mexico desert, near the Elephant Butte Dam, which was built to control the Rio Grande, and the home of outlaw Billy the Kid. As well as catching up on the area's historic past, those on the tour are allowed inside the spaceport perimeter, and can view the airfield, air rescue facilities, fuel storage complex, and the iconic terminal hangar facility. The odd surprise may be in store, such as a passing cow, since the spaceport shares the land with two working cattle ranches. The tour also includes updates on the commercial space industry. The Spaceport America Preview Tour lasts between three and four hours, costs US$99 and can be booked through Follow the Sun Tours (*www.ftstours.com*).

for Virgin Galactic isn't the only spaceport being built. In fact, Spaceport America is just one of more than half a dozen spaceports worldwide, the others being Aeroports de Catalunya, Cecil Field Spaceport, Mojave Air and Space Port, Oklahoma Spaceport, Space Florida, Spaceport Indiana, Spaceport Scotland, Spaceport Sweden, and Wisconsin Aerospace Authority (the newest member of the club is Space Experience Curacao, whose first customers are expected to be companies selling rides in XCOR's Lynx). The spaceports are all members of the Spaceports Council, which operates under the aegis of the Commercial Spaceflight Federation (CSF).

ORGANIZATIONS

Just like the government has NASA, commercial spaceflight has its own sort of agency, called the Commercial Spaceflight Federation (CSF). The CSF is a private spaceflight industry group, incorporated as an industry association for the purposes of establishing safety standards for the commercial human spaceflight industry, sharing best practices and expertise, and promoting the growth of the industry. The sort of issues the CSF work on include initiatives such as searching for grants programs for spaceport infrastructure, dealing with FAA regulations and permits, industry safety standards, and, of course, public advocacy for the commercial space sector. The CSF was initially conceived as the Personal Spaceflight Federation (PSF) in 2005. Back then, the goals of the PSF included member coordination, government interface, safety in spaceport operations, crew and passenger training, vehicle manufacture, operations, insurance, and public relations. Then, in June 2008, the PSF changed its name to the CSF and it refocused its goals, choosing to focus on cargo and crew to the ISS, the flight of private individuals, science research missions, R&D, astronaut training, and education and outreach activities. In addition to being instrumental in the creation of the Spaceports Council, the CSF also played a part in the creation of the Suborbital Applications Research Group (SARG). SARG, a coordination and advisory committee of the CSF, is composed of scientists and researchers dedicated to furthering the scientific potential of suborbital reusable launch vehicles under development by the commercial spaceflight sector. The committee's main focus is to increase awareness of commercial suborbital vehicles in science, R&D, and working with policymakers to ensure that payloads have easy access to these vehicles. The committee is chaired by Dr Alan Stern of the Southwest Research Institute (SwRI), which has already committed funding to fly researcher-astronauts and their payloads on board commercial suborbital spacecraft. SwRI also played a leading role in kicking off the Suborbital Scientist Training Program (SSTP) in conjunction with NASTAR and played a pivotal role in organizing the Next-Generation Suborbital Researchers Conference (NSRC), which was discussed in Chapter 1.

REGULATIONS

Commercial spaceflight has no problem stealing headlines across the news media. And why shouldn't it? With a number of commercial companies ramping up to operational status, progress is swift, and any story about a new spacecraft for which we'll someday be able to buy a ticket into space makes for great press. Less sexy, but no less important, is the regulatory structure that will make the commercial space flight revolution possible.

The creation of the commercial space flight sector can be traced back to the passage of the Commercial Space Launch Act (CSLA) in 1984. The CLSA laid out guidelines for the acquisition of launch licenses for private rockets, and granted the Office of Commercial Space Transportation (OCST) the authority to grant them. Today, the OCST is part of the FAA, which is the agency responsible for coalescing the regulations for reusable craft and rules for their human passengers. Recently, the FAA partnered with a handful of US academic institutions to work on the nuts and bolts of commercial space regulations. The outcome was the creation of the Center of Excellence for Commercial Space Transportation, which was established by Congress in 2010. The center's efforts will focus on four major research areas: space launch operations and traffic management; launch vehicle systems, payloads, technologies, and operations; commercial human space flight; and space commerce.

Now, you may be wondering why there's a need for so many regulations; after all, space is extra-territorial. And won't all these regulations just slow the development of commercial spaceflight, especially if aviation-like regulatory safety is implemented? Well, bureaucracy is always slow, but international law makes nation states liable for the activities of its citizens and regulations are required by each country to ensure responsible behavior while reducing potential liability. Let's face it, spaceflight is unlike any other transportation mode, and its dismal safety record is on a par with climbing Mt Everest and several orders of magnitude worse than extreme sports like skydiving. In the era of commercial enterprises taking astronauts for hire on suborbital flights, the safety and comfort of the passengers will be paramount; after all, there will be no business if the vehicles can't get up to space and back in one piece!

So, safety has to be the first concern; only by distributing best practices while working towards voluntary industry standards in spaceport operations, vehicle manufacture, and crew and passenger training will the agency ensure safety issues are acknowledged and addressed by the launch providers. To that end, the FAA has issued a proposed rule, entitled "Human Space Flight Requirements for Crew and Space Flight Participants", in which the FAA recognizes that commercial spaceflight is an industry in evolution, and doesn't wish to constrain the growth or innovation of the participants. In this proposed rule, the FAA makes several clear statements about the training and preparation for crew and passengers that launch providers should strongly consider. These statements include training for abort scenarios and emergency conditions as well as demonstrating the ability to withstand the stresses of spaceflight.

Although the ideal scenario is to train crew and passengers in the actual vehicles and situations they will encounter, this is not currently practical, since, in most cases, these vehicles and situations don't exist. Several alternatives are under consideration. Some suborbital vehicle builders believe training in stunt planes should be adequate for the task, while others, such as Astronauts for Hire (A4H), have developed a formal generic suborbital training program (see Chapter 1).

Another aspect of training private space travelers is preparing them for the experience; after all, a suborbital launch will be very different from boarding a commercial airline flight. Passengers can expect to experience anywhere from 4 to 7 G on launch and re-entry, along with an assortment of discombobulating stimuli. If such stresses are encountered for the first time during a launch, or if the passenger is inadequately trained, they may experience several adverse effects, such as vomiting, making the whole event rather unpleasant – an experience that might not reflect favorably on the launch provider!

One of the keys to avoiding an unpleasant flight is simulation, a training element the FAA explicitly acknowledges. That's the reason Virgin Galactic trains their spaceflight participants at the NASTAR Center, which can provide realistic flight models of practically any spacecraft, thereby providing as realistic a launch and re-entry experience as possible, short of the actual flight.

While the FAA seem to be taking a lenient approach to certification and training regulations, Europe's policy may force spacecraft operators to operate in a very different regulatory environment. That's because the European Aviation Safety Administration (EASA) plans to certify winged vehicles that will fly into space under its authority to regulate aircraft. According to EASA's regulations, an aircraft is "any machine that can derive support in the atmosphere from the reactions of the air other than the reactions of the air against the earth's surface". By this definition, suborbital airplanes, which generate aerodynamic lift during the atmospheric part of their flights, are considered to be aircraft. Of course, certifying aircraft is an expensive and time-consuming process that many commercial space operators would like to avoid. The FAA's Commercial Space Transportation Advisory Committee (COMSTAC) agrees, which is why it has adopted a "learn as you go" approach. COMSTAC reckons that imposing aircraft-like certification of winged space vehicles could hurt the early development of the suborbital space transportation industry but, if EASA's proposal goes ahead, what impact would that have on commercial space operators?

To try to breach the gap, the FAA is promoting the concept of "interoperability" versus "harmonization" for spaceflight regulations. The FAA defines interoperability as "the ability to operate a vehicle in multiple regulatory regimes without major changes". Basically, what this means is, if a US company wants to launch a US-operated commercial vehicle abroad, it would require an FAA launch license and would need to comply with foreign safety regulations, which would probably require some kind of memorandum of understanding. However, this might prove tricky because of US export policy restrictions on space technology transfer, which means the US can't cooperate in space transportation to the level it can with aviation.

How will the COMSTAC/EASA difference be resolved? Well, harmonization could be achieved as a result of a trigger event such as an operational vehicle wanting to operate in two different regulatory regimes or an orbital re-entry to a country that's different from the launch site country. It will be interesting to see which approach works best. Time will tell.

5

The New Breed of Space Vehicles

Fifty years after Russian cosmonaut Yuri Gagarin became the first human to experience the thrill of the final frontier, commercial companies are on the cusp of another manned spaceflight milestone – a revolutionary breakthrough industry that will open up suborbital and orbital space to commercial astronauts. Suborbital spaceships will take astronauts for hire into space at an altitude of 100 km before returning to Earth. On orbital flights, this new cadre of astronauts will ride into low-Earth orbit (LEO) and perhaps visit one of Bigelow's inflatable habitats. In the meantime, commercial spaceflight seems poised to launch, and there are a wide range of options available; let's start with the suborbital operators.

THE NEW AREA 51

Mojave, California: it's a low-rise town of 3,800 people, about 160 km north of Los Angeles and a short drive from Edwards Air Force Base and the Navy's biggest weapons test facility at China Lake. Desert winds blow relentlessly down the dusty main drag, past the cluster of fast-food franchises and the sole halfway decent roadhouse, Mike's, where you'll most likely find miners, bikers, and pilots either drinking or shooting pool. Walking around the non-descript town, it's hard to imagine that this is ground zero for the commercial spaceflight industry, but aerospace history is often made in isolated places where risky designs and radical procedures can be tested in places where mistakes won't crash into subdivisions or strip malls. Before Mojave, perhaps the most famous test ground was the US Air Force base at Groom Lake, Nevada, a secret facility that inspired iconic nicknames like Dreamland and Area 51. For all the alien-conspiracy hype, Area 51 legitimately secured its place in history as the birthplace of revolutionary aircraft such as the high-flying U-2 spy plane and the supersonic SR-71 Blackbird; at its zenith during the Cold War, Area 51 was a top-secret site where defense contractors worked on all manner of classified aircraft. "Black" projects are alive and well in Mojave, as attested by Jim Balentine, president of the airport district's board of directors, who admits that more often than not, the tenants aren't too keen to disclose who they are or what they're working on.

However, Mojave qualifies as the new Area 51, not because of large military contractors, but because it has lured entrepreneurial commercial space players. You see, for a company whose goal it is to design and develop a new generation of spacecraft, Mojave is the perfect location: no homeowners for miles, super-long runways, access to reserved military airspace, and a supersonic flight corridor, thanks to an agreement with Edwards Air Force Base, 30 km down the road. And, if you need a hot-shot pilot, then the world-renowned National Test Pilot School (NTPS) provides a steady stream of cockpit jockeys. All this helps to make Mojave *the* place to go for designers with experimental aircraft to test.

Mojave's first airport, built in 1935, originally served the surrounding gold and silver mines. When the government appropriated it during the Second World War as a Marine auxiliary air station, Corps pilots used it as a place to practice their gunnery. Then, when the Marines pulled out in 1961, the airfield might have faded back into the desert if not for Dan Sabovich, an aviation-obsessed rancher. Sabovich, who liked to fly his V-tail Beechcraft Bonanza from his own strip on his spread near Bakersfield, imagined Mojave as a civilian test-flight center that would cater to experimental aviation and be run by its own elected officials who could protect its spirit of adventure. In 1972, thanks to Sabovich's political acumen, and persistent intense lobbying, the state created Mojave's special airport district, which Sabovich ran until 2002. Shortly after the airport district was founded, Burt Rutan, a young aeronautical engineer from California Polytechnic, set up the Rutan Aircraft Factory – a move that was followed by other aviation companies, although it wasn't until the 1990s that the first rocketeers began to arrive. One of the first was Rotary Rocket, a company that had designs on developing a low-cost, manned, reusable spacecraft, Roton. The unconventional vehicle was designed to be launched like a conventional rocket and boosted by a novel rotary engine that burned kerosene and liquid oxygen. After re-entry, pilots were supposed to make helicopter-style landings using nose-mounted rotors, but the company never really got off the ground (excuse the pun!) and folded in 2001. Three years later, the FAA certified Mojave as a spaceport, which means private firms can launch craft into orbit. Stuart Witt, the current general manager of the airport, continues to preserve Sabovich's mission, by offering cut-rate rents to start-up companies.

Designing and developing new spacecraft is a risky business. In July 2007, at the rocket test range, Scaled Composites conducted a cold-flow test on a new engine component for SpaceShipTwo (SS2). Three seconds into the test, a pressurized tank of nitrous oxide exploded, killing three people and injuring three others. A six-month investigation ensued, led by the FAA and California work safety regulators. After federal and state inspectors were unable to determine the exact cause of the accident, the FAA threatened to withdraw Mojave's spaceport license but, after inspecting the facilities, had second thoughts and instituted safety-related amendments to the license.

Wander over to the northern edge of town and you'll find a chain-link fence that marks the boundary of the Mojave Air and Space Port. The spaceport (Figure 5.1), which sprawls across 3,300 acres of the Mojave Desert, features a control tower that stands sentinel over three runways, the longest of which extends more than 3 km out into the scrubby flats. Weather-beaten hangars, some dating back to the Second World War, line the main runway. Venture inside one of these hangars and you'll

Figure 5.1 Mojave Spaceport. Courtesy: Mojave Spaceport

understand why Mojave is the focal point of aerospace research and development. Secret Pentagon programs, black projects, implausible-looking aircraft, and, of course, private spaceships – they all take shape in these aluminum-sided buildings. For obvious reasons, most hangar doors are shut tight, but the few that are open offer glimpses of gleaming pressurized tanks, technicians in oil-smeared overalls, and smooth pearlescent fuselages, usually labeled with a distinctive black insignia mandated by the FAA: Experimental.

THE X-15 LEGACY

To the east of Mojave is Edwards Air Force Base, home for legendary pilots and the milestone-making NASA X-15 rocket plane (Figure 5.2), which roared to life high above the Mojave landscape for nearly a decade (the program included 199 flights between 1959 and 1968). Flying close to the edge of space, and sometimes beyond, the X-15 half-plane, half-rocket flights mimicked SpaceShipOne's (SS1's) private trek to space more than three decades later.

The X-15 program was a joint effort of NASA, the US Air Force (USAF), and the US Navy (USN). During the program, eight pilots[1] met the US criterion for

---

[1] Robert White, Joseph Walker, Robert Rushworth, John "Jack" McKay, Joseph Engle, William "Pete" Knight, William Dana, and Michael Adams.

Figure 5.2 As crew members secure the X-15 rocket-powered aircraft after a research flight in September 1961, the B-52 mother-ship used for launching this unique aircraft does a low fly-by overhead. NASA B-52, Tail Number 008, served as the launch vehicle on 106 X-15 flights and flew 159 captive-carry and launch missions in support of the program from June 1959 to October 1968. A wing pylon installed on the right wing, between the inboard engine pylon and the fuselage, enables the B-52 to carry research vehicles and test articles to be air-launched/dropped. Courtesy: US Air Force

spaceflights by passing an altitude of 80 km and were accordingly awarded astronaut status by the USAF. Two of the X-15 pilots also qualified for recognition by the Fédération Aéronautique Internationale (FAI). After its initial test flights in 1959, the North American-built X-15 became the first winged aircraft to reach hypersonic velocities of Mach 4, 5, and 6 and to operate at altitudes above 30,500 m. In common with SS2, the X-15 was carried under the wing of another aircraft (Figure 5.3), in this case a Boeing B-52 bomber – Scaled Composites built their own B-52, which they call WhiteKnight.

The wedge-shaped tail surfaces of the X-15 provided directional stability at speeds at which conventionally shaped airfoils would have had little effect. The large upper and lower fins and the downward slant of the wings allowed the X-15 to remain stable during steep climbs and at high altitudes. Covering the titanium substructure was a layer of Inconel X, a nickel alloy capable of withstanding temperatures of 650°C (more heat dissipation was achieved by coloring the fuselage black). Not surprisingly, Inconel X, like the airframe, thermal protection, flight controls, aerodynamic performance, pilot protection, and just about every other feature on the X-15, was experimental. For example, the aircraft included the first inertial navigation system (INS) and used a sophisticated Stability Augmentation System

Figure 5.3 This photo illustrates how the X-15 rocket-powered aircraft was taken aloft under the wing of a B-52. Because of the large fuel consumption, the X-15 was air launched from a B-52 aircraft at 45,000 ft and a speed of about 500 mph. This photo was taken from one of the observation windows in the B-52 shortly before dropping the X-15. Courtesy: US Air Force

that helped maintain control of the aircraft. Above the atmosphere, reaction controls kept the aircraft stable, while hand controllers in the cockpit were linked to small hydrogen peroxide thrusters at the nose for pitch and yaw control, and on each outer wing to control roll. During the descent, the pilot switched from thruster control to traditional stick-and-rudder flying, effectively making the X-15 a 15,000-lb unpowered glider (Figure 5.4), much like its direct descendant, SS2.

One feature SS1 didn't have in common, despite its being an experimental vehicle, was an ejection seat. Unlike the SS1 pilots, who have no ejection seat (and no high-performance flight suit, incidentally) and who have to hope everything works perfectly, the X-15 pilots could avail themselves of a seat that enabled ejection at speeds up to Mach 4, in any attitude and at any altitude up to 120,000 ft. After the canopy was jettisoned, the seat fired upward and, once clear of the X-15, deployed a pair of fold-out fins and telescoping booms for stabilization. Oxygen was supplied from two under-seat tanks and restraints kept the pilot's arms and legs from flailing in the airstream. After reaching a safe altitude, the pilot would unbuckle, jump from the seat, and activate his parachute. Powering the X-15[2] was the liquid oxygen and anhydrous ammonia-fuelled XLR99 rocket engine that generated an awesome half a

---

[2] SS1 was fuelled by nitrous oxide, more commonly known as "laughing gas", which is what dentists used to use as anesthesia, and rubber.

Figure 5.4 X-15. Courtesy: US Air Force

million horsepower (another unique feature of the X-15 was its engine, which was restartable).

A typical X-15 flight started with the ground crew disconnecting the servicing carts used to prepare the B-52 and the X-15 for flight. The B-52 then taxied along the dry lake bed before take-off; on a hot day, most of the 12,000 ft of runway was required to achieve liftoff. Twelve minutes prior to launch, the pilot started the auxiliary power units, checked his systems and flight controls, tested the ballistic control system, set the switch positions, activated the propulsion system, and turned on data recorders and cameras. With his preflight checks completed, the pilot prepared for release from the B-52 – a gut-wrenching change from normal 1 G to zero-G flight. A few seconds of freefall followed, and then the powerful rocket engine ignited and catapulted the X-15 to Mach 5 and beyond. Climbing at an incredible 4,000 ft per second, the engine shut down after one or two minutes, leaving the X-15 as a super-glider. From sheer momentum, the aircraft continued to climb before orienting itself for an unpowered landing. During a typical descent, the pilot would be subject to between 5 and 8 G during re-entry, which is similar to the 6 G SS2 passengers will experience. At 35,000 ft, the pilot set up his final approach for landing at Edwards Air Force Base, dumping any remaining propellant to reduce weight, before jettisoning the ventral rudder from the aft underbelly and lowering the landing flaps, nose gear, and rear landing skids. Touchdown would normally be made at a sink rate of two feet per second and a forward velocity of 200 miles/hr with a slide-out distance of up to 6,000 ft.

The X-15 was a pure research vehicle, whose sole function was to explore the effects of hypersonic travel on man and machine. In the end, the X-15 program yielded a treasure trove of hypersonic lessons learned that proved invaluable to developers of the Space Shuttle and, more recently, to the commercial space vehicle developers such as Scaled Composites.

SCALED COMPOSITES

The hangar that has attracted the most media attention in Mojave is one owned by Scaled Composites, the company responsible for the winning spacecraft entry of the Ansari X PRIZE. The space craft, known as SpaceShipOne, achieved the set goal of the competition on two separate occasions and Scaled Composites won the $10 million prize offered by the X PRIZE competition. The goal of the competition was to foster private research and development of space flight and the competition went on to draw over $1 billion in private investments, with Scaled Composites being one of the companies to benefit. Following their X PRIZE success, Scaled Composites searched for a way to commercialize their achievement – a move that led to its partnership with Virgin CEO, Sir Richard Branson, to create one of the first private space flight companies: Virgin Galactic. The company was contracted to build several second-generation space craft that would be used for suborbital spaceflights. That was in 2004. Seven years later, under the guidance of mutton-chopped aerospace samurai, Burt Rutan, the company has emerged as one of the major players in the nascent commercial spaceflight industry.

With SpaceShipOne safely berthed in the Smithsonian museum, Rutan's Scaled Composites engineering crew locked down the designs and started laying down carbon fiber for two new craft: a supersize SS2 (Figure 5.5) and a reconceived WhiteKnightTwo (WK2) mother ship to carry it aloft.

Like its X PRIZE-winning predecessor, SS2 will carry passengers to the edge of space at 5,000 km/hr. But, while the original SS2 was barely a three-seater, the roomier version (Figure 5.6) is designed for six passengers and two pilots. New features such as bigger windows and a larger cabin are designed to maximize the "wow" factor of being in space. But, while SS2 – now known as VSS Enterprise – is simply a bigger version of the original rocket, the jet-powered WK2 is nothing short of a total redesign. The first mother ship had a single fuselage. The new one accommodates the expanded ship with twin hulls that hold the VSS Enterprise suspended between them. In the summer of 2011, VSS Enterprise was still very much a work in progress; the interior was lacking even seats, although the fittings where the rotating passenger seats will be installed were fitted. Although the cabin size is relatively modest, it will probably look a bit cramped when fully outfitted and loaded with six passengers, which may mean some bumping and colliding during the

118 The New Breed of Space Vehicles

Figure 5.5 SpaceShipTwo. Courtesy: The SpaceShip Company

Figure 5.6 SpaceShipTwo interior. Courtesy: The Spaceship Company

weightless portion of the flight – just like commercial airline travel, come to think of it!

Since its unveiling, VSS Enterprise has racked up milestone after milestone; in March 2010, Virgin Galactic conducted the vehicle's inaugural "captive carry" test flight when VSS Enterprise spent 2 hours 54 minutes attached to Eve, the WK2 carrier aircraft. The captive carry test achieved an altitude of 13,716 m, which is just a little lower than the release altitude planned for revenue flights. The next flight-testing milestone occurred in October 2010, when the VSS Enterprise made its first manned glide flight. The spacecraft was released from Eve at 13,700 m and flew, unpowered, to a landing at the Mojave Spaceport. Scaled Composites' test pilot Pete Siebold pronounced the VSS Enterprise "A real joy to fly, especially when one considers the fact that the vehicle has been designed not only to be a Mach 3.5 spaceship capable of going into space but also one of the worlds' highest altitude gliders".

During the flight, all systems operated trouble free and, after a clean release, Siebold completed initial flight handling and stall characteristic evaluation of VSS Enterprise before completing a practice approach and landing, followed by a smooth landing. The manned glide flights continued throughout the rest of 2010 and into 2011. The fifth, and longest, flight occurred in April 2011, when test pilot Peter Siebold and co-pilot Doug Shane, president of Scaled Composites, flew VSS Enterprise for 14 minutes 31 seconds. This milestone flight was followed by another landmark the following month when the vehicle demonstrated its unique re-entry "feather" technique.[3] The flight, the spacecraft's seventh since its rollout in December 2009, climbed to its usual altitude before being released. It then established a stable glide profile before deploying its re-entry "feathered" configuration by rotating the tail section of the vehicle upwards to a 65° angle to the fuselage (Figure 5.7). It remained in this position for more than a minute whilst descending almost vertically at 15,500 ft per minute, steadily being slowed by the shuttlecock-like drag created by the raised tail section. At 10,000 m, the pilots established VSS Enterprise's normal glide mode and performed a smooth landing, bringing to an end yet another successful test flight and reinforcing the fast turnaround – the flight was VSS Enterprise's third in 12 days – and frequent flight-rate potential of the vehicle.

---

[3] In the feather position, SS2's tail is folded 65° upwards with reference to the fuselage – a configuration allowing the spacecraft to descend nearly vertically back into the atmosphere while maintaining an airspeed slow enough to prevent extreme heating on the fuselage. The feathered phase of the flight is designed to be simpler than flying the spaceship back into the atmosphere, as was done by the X-15 back in the 1960s. However, the ride is still a very dynamic one, with SS2 descending like a shuttlecock with swaying and pendulum motion.

120 The New Breed of Space Vehicles

Figure 5.7 SpaceShipTwo feathering. Courtesy: The Spaceship Company

XCOR

Just down the road from Scaled is a faded blue hangar housing another suborbital start-up. XCOR, a Mojave-based outfit that invented a helium-powered rocket-fuel pump for a Defense Advanced Research Projects Agency (DARPA) project, is developing its Lynx spaceplane. It won't reach as high or carry as many passengers as VSS Enterprise, but the tickets will be cheaper – about $95,000.

Although the stubby-nosed Lynx will be a sub-suborbital vehicle (initial flights will be below 100 km), XCOR has its sights on the bigger prize of suborbital and eventually orbital travel. You'd think that a company building a spaceship would be pretty sizeable, but that's not the case with XCOR, which has fewer than 50 employees on the payroll. But, despite their small size, the company has developed no fewer than 11 different rocket engines, and three different rocket vehicles!

There are some in the media who have incorrectly labeled XCOR as Virgin Galactic's closest competition, which is unfair because it implies they offer a similar service to Branson's company. The reality is that XCOR offers a shorter flight, to a sub-space altitude, on a less powerful vehicle that allows passengers less time to float around – which will be almost impossible anyway due to the tight confines of the Lynx. Also, Lynx passengers have to wear a spacesuit, while Virgin Galactic customers are relatively unencumbered thanks to their snappy flight suit with visor. And, finally, XCOR's passenger gets to view whatever there is to view at 200,000 ft through the

Lynx's wraparound cockpit while their Virgin Galactic counterparts spend four or five minutes floating around the capacious cabin. Basically, Virgin Galactic's VSS Enterprise will fly you into space while XCOR will give you an X-15 experience: a pilot's eye view of Earth from 40 miles' altitude followed by a gripping 4-G descent. Will it be enough to keep them in business? Perhaps not, which is why XCOR is taking a step-by-step approach to their ultimate goal of reaching orbit, which is probably where the big money is. But, don't forget, revenue flights won't be just limited to the well-heeled tourist; astronauts for hire, representing the space science community, will no doubt fly on sub-suborbital vehicles like the Lynx, which is why XCOR has designed the vehicle to serve as much of that market as possible. The draw of the Lynx is that it offers scientists the chance to fly frequently, on demand, and at a low price. Rather than waiting months for a sounding rocket to become available, researchers, and the astronauts for hire representing them, will be able to fly their payloads on Lynx multiple times per week. While the flights won't offer the cache of having flown in orbit, it may prove a useful income stepping-stone on the way to orbit.

The Lynx is a two-seated reusable launch vehicle (RLV) designed to take a pilot and passenger on a half-hour suborbital flight to 100 km and return to a landing at the take-off runway. Like an aircraft, Lynx is a horizontal take-off and horizontal landing vehicle, but instead of a jet engine, Lynx uses a reusable rocket propulsion system to depart a runway and return. It's an approach that's unique compared to most other RLVs in development, such as conventional vertical rocket launches and air-launched winged rocket vehicles like VSS Enterprise.

Structurally, the Lynx features an all-composite airframe (Figure 5.8) that makes it lightweight and strong, and, thanks to a thermal protection system (TPS) on the nose and leading edges, the vehicle is able to handle the heat of re-entry from the edge of space. At just 9 m in length with a double-delta wing that spans about 7.5 m, the Lynx is the space equivalent of the Smart Car.

Once operational, the Lynx will operate as an FAA AST-licensed suborbital reusable launch vehicle, taking off up to four times a day from any licensed spaceport with a 2,400-m runway. Unlike VSS Enterprise, the Lynx is configured more with science than tourism in mind; thanks to its dorsal payload pod, the vehicle will offer multi-mission primary and secondary capabilities, including in-cockpit experiments, externally mounted experiments, test pilot/astronaut training, microsatellite launch, and even ballistic trajectory research. Primary payloads (Figure 5.8) will be carried in the area to the right of the pilot or on the top of the vehicle in an experiment pod, while secondary payloads will be carried in a variety of locations. For example, the production version of the Lynx will accommodate a maximum primary payload mass of 120 kg to 100 km (330,000 ft), which can either be a human in a pressure suit or a standard electronic equipment rack. The other primary payload location is an external dorsal mounted pod (Figure 5.9), which holds up to 650 kg and is large enough to hold a two-stage carrier to launch a microsatellite or nanosatellites into LEO. This pod can also carry experiments that return with the vehicle.

When it isn't deploying nanosatellites or ferrying equipment racks into space, the Lynx will carry a passenger, who will sit to the right and just aft of the pilot. The Lynx will have a pressurized cabin, but the pilot and participant will still wear full

122 The New Breed of Space Vehicles

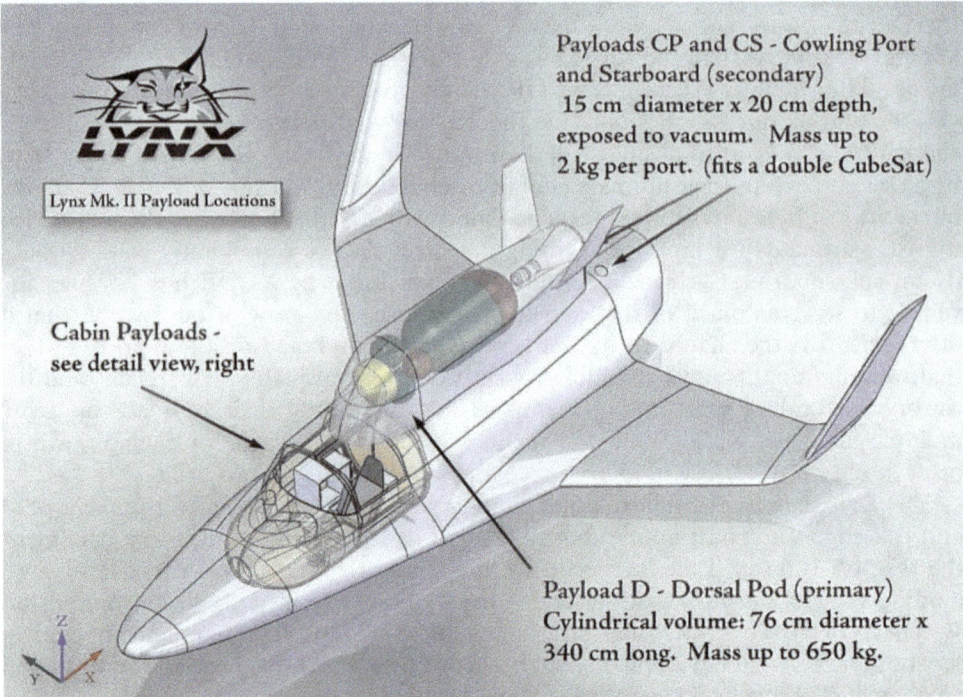

Figure 5.8 Infographic showing the interior of the Lynx. Courtesy: XCOR

Figure 5.9 Lynx fitted with cargo pod. Courtesy: XCOR

pressure suits (designed by Orbital Outfitters) – just in case. Before their flight, passengers will have undergone a screening process, followed by five days of training that will include seminars, altitude chamber training, and a G-force experience.

DREAM CHASER

Far away from the Mojave Desert, in the basement of the University of Colorado's aerospace engineering department, another suborbital enterprise is taking shape. Here, a team of graduate students sit in a fiberglass cockpit, simulating flipping switches on a dashboard, divided into sections, with control switches and color printouts of monitor displays. Surrounding the cockpit are NASA spacesuits, helmets, gloves, boots, and other space paraphernalia. The students are working on human-factor assessments of a cockpit for Sierra Nevada Corporation's new spaceplane – Dream Chaser – that will transport cargo and crew to the International Space Station (ISS). Thanks to $80 million of NASA funds, Sierra Nevada is hoping to fill the gap left by the retired Shuttle by building a 10-m-long space plane (Figure 5.10), capable of carrying seven crew members to LEO.

Helping the graduate students with their project is Jim Voss, a retired astronaut working with Sierra Nevada and the university. It's a win–win partnership: Sierra Nevada funds the university project and, in return, has access to university resources, including the eager students who get real-world experience. The partnership also allows the company to design their spacecraft cost-effectively, while training aerospace engineers (two graduate students have already been hired by Sierra Nevada). Unlike VSS Enterprise's distinctive, one-of-a-kind design, Dream Chaser bears more than a passing resemblance to the retired Shuttle, albeit a scaled-down version – although, despite its smaller size, it can still carry up to eight people. However, the business jet-sized spacecraft does share some technology with VSS Enterprise: its hybrid motors, which burn an unusual fuel – a combination of recycled rubber and nitrous oxide (this has almost the same energy density as conventional fuel but can be burned in a more controlled way, which might be safer, although only multiple flights can prove it).

Unlike the Shuttle, which launched into orbit thanks to two strap-on solid-rocket boosters and a huge tank containing hypergolic fuels, Dream Chaser – which is a derivative of NASA Langley's HL-20[4] (Figure 5.11) – will launch atop a rocket,

---

[4] As a result of the Space Shuttle *Challenger* accident, there was renewed interest in developing a crew emergency rescue vehicle (CERV) for the ISS in the event the Shuttle was unavailable or if astronauts had to return to Earth in an emergency. In late 1986, the Vehicle Analysis Branch (VAB) began studying the use of the lifting body shape as a CERV big enough to accommodate up to eight ISS crewmembers. In October 1989, Rockwell International (Space Systems Division) began a year-long effort managed by Langley Research Center (LRC) to perform a study of the Personnel Launch System (PLS) design and operations with the HL-20 as a baseline. In 1991, a full-size engineering research model of the HL-20 was constructed by the students of North Carolina State University. The program was cancelled in 1993.

124 The New Breed of Space Vehicles

Figure 5.10 Sierra Nevada's Dream Chaser. Courtesy: Sierra Nevada

Figure 5.11 NASA's HL-20 lifting body mockup was a full-scale non-flying mockup used for engineering studies, testing crew positions and pilot visibility. Courtesy: NASA

either a suborbital hybrid booster or an Atlas V. In common with the Lynx, Dream Chaser can be adapted to various mission configurations, including carrying all passengers, pressurized or unpressurized cargo, or various combinations of crew and pressurized cargo spacecraft. If the company continues to achieve its testing and development milestones, the Dream Chaser will be launched into orbit in 2014 on the nose of an Atlas V, made by United Launch Alliance. Once it reaches orbit, the vehicle will be dropped, and its hybrid motors will be used to adjust its orbit or dock it to the ISS. These motors will also be used to guide it to its landings.

While NASA is the company's primary customer, Sierra Nevada expects that universities and research institutes will be interested in buying space on the Dream Chaser to send experiments into orbit. What the cost of flying these experiments into orbit will be is something the company hasn't disclosed, but Sierra Nevada is a profitable company, so the price point will be one that ensures a return on their investment. Fortunately, everything on the Dream Chaser except the initial launch booster and the fuel cartridges is designed to be reused, so turning a profit on the tens of millions of dollars the company has invested in the project may be achievable.

126   The New Breed of Space Vehicles

BLUE ORIGIN

Blue Origin's New Shepard (a reference to the first American astronaut in space, Alan Shepard) RLV is a vertical-take-off and landing (VTOL), suborbital manned rocket powered by high test peroxide (HTP) and RP-1 kerosene. The New Shepard vehicle (Figure 5.12) will consist of a pressurized Crew Capsule (CC) carrying experiments and astronauts atop a Propulsion Module (PM). Flights will take place from Blue Origin's launch site, in West Texas. New Shepard will take-off vertically and accelerate for approximately two-and-a-half minutes before shutting off its rocket engines and coasting into space. The vehicle will carry rocket motors enabling the CC to escape from the PM in the event of a malfunction during launch. Once in space, the CC will separate from the PM and the two will re-enter and land separately (the CC will land under a parachute) for reuse. Astronauts and experiments will be subject to no more than 6-G acceleration and 1.5-G lateral acceleration during a typical flight, which will provide a high-quality microgravity environment. With its eye on science as a revenue-generator, Blue Origin is already soliciting input from investigators to help design experiment accommodations. Researchers will have the opportunity to provide their own racks to mount into the vehicle, or use standard racks and services to mount their experiments. In addition to

Figure 5.12 Blue Origin's orbital spaceflight in-flight rendering. Courtesy: Blue Origin

in-cabin science, Blue Origin also plans to provide researchers with the capability of conducting remote sensing, such as atmospheric science and Earth observations, sampling of the atmosphere, and magnetospheric measurements. This interest in suborbital science (dubbed research and education missions – REMs) is fuelled by NASA's creation of a program office dedicated to explore this arena at the space agency's Ames Research Center (ARC).

Funded by NASA's CCDev-2, Blue Origin's biconic New Shepard (aka, the Space Vehicle) will be capable of carrying seven astronauts and will transfer NASA crew and cargo to and from the ISS, serve as an ISS emergency escape vehicle for up to 210 days, and perform a land landing, thereby minimizing costs of recovery. Although the Space Vehicle is designed to ride on multiple boosters, the Atlas V was chosen for initial capability thanks to its proven launch track record and performance capability. In tandem with its space vehicle development, Blue Origin is developing its Reusable Booster System (RBS) to lower the cost of space access. To achieve the VTOL capability, the RBS employs deep-throttling, restartable engines. Rated at 100,000 lbf, the liquid oxygen/liquid hydrogen engines are pacing the entire orbital RBS program. Thanks to the company's $22 million CCDev-2 funding, the company is working on maturing its orbital space vehicle design by completing key system trades, including design of the thermal protection system, definition of the biconic shape including aerodynamic analyses, CFD analysis, and wind tunnel testing. In addition to working on the system trades, Blue Origin is also developing a pusher escape system as a part of its space vehicle. The system, which permits a crew escape capability in the event of an anomaly on the launch vehicle, uses an engine in a pusher configuration, which is substantially different from the traditional towed-tractor escape tower concepts utilized on Mercury, Apollo, and Orion.

BIGELOW AEROSPACE

Just over 10 years ago, Robert T. Bigelow committed $500 million of his own money to realize his vision of creating a commercially developed space station and a Moon base. He licensed a canceled NASA project called Trans-hab, added 14 of his own patents, and created an expandable habitat. One (Genesis I) was launched in 2006 and another (Genesis II) in 2007. They're still up there, working flawlessly. In fact, Bigelow Aerospace was so pleased with Genesis I and II that they skipped over an interim craft to develop even bigger habitats, one of them three times the volume of the average module on the ISS. The two commercial habitats – the 180-cubic-meter Sundancer and the larger BA-330 (Figure 5.13) – will be launched in 2014 and 2015.

The BA-330 (previously known as the Nautilus space complex module) is a full-scale space habitation module with 330 cubic meters of internal space. Once deployed in LEO, the habitat will offer a large habitable space for crews to live and conduct experiments in. Weighing just 20,000 kg, the rugged BA-330 (Panel 5.1) is made of high-strength textiles and Vectran-like materials, wrapped with several layers of high-tension straps. While the term "inflatable" may have some questioning

128 The New Breed of Space Vehicles

Figure 5.13 Bigelow Aerospace BA-330 modules. Courtesy: Bigelow Aerospace

the habitat's strength (one of my A4H colleagues who worked for Bigelow Aerospace had the opportunity to touch the habitat and said it was hard as concrete), the habitat is particularly resistant to damage from micrometeoroids and space debris and provides radiation protection equivalent, and ballistic protection superior, to the ISS. Just like its government-owned counterpart, the BA-330 will be self-sufficient, its electrical power provided by solar arrays and batteries, avionics provided for navigation, re-boost, docking, and other on-orbit maneuvering, a fully fledged environmental control and life-support system, and four large windows to support both terrestrial and celestial viewing.

Sundancer will be launched first to test its capabilities and, provided the testing phase goes well, a second Sundancer and a docking node bus will be deployed by 2015, followed by a BA-330. Once in orbit, the habitats will be populated by well-heeled sovereign customers, governments, and corporations. The company is pitching Bigelow's space station as an affordable alternative to the ISS – a platform that can take advantage of an untapped market in companies that would use a space platform to perform materials work, undertake pharmaceutical research, or conduct experiments in microgravity. To launch the habitats into orbit, the company is eyeing United Launch Alliance's Atlas V and SpaceX's Falcon-9 rockets as candidates but, with launch capacity limited, Bigelow may have a tough time securing all the services it will need to orbit its space station, which is why Bigelow is hoping for a viable crew transfer vehicle (CTV) being ready by 2014. The company is

> **Panel 5.1. BA-330**
>
> The BA-330's skin is composed of almost two dozen layers, fashioned to break up particles of space debris and micrometeorites that may hit the shell with a speed seven times as fast as a bullet. The outer layers protect multiple inner bladders, made of a material that holds in the module's air. The key to the debris protection is successive layers of Nextel, a material commonly used as insulation under the hoods of many cars, spaced between several-inches-thick layers of open cell foam, similar to foam used for chair cushions on Earth. The Nextel and foam layers cause a particle to shatter as it hits, losing more and more of its energy as it penetrates deeper. Many layers into the shell is a layer of super-strong woven Kevlar that holds the module's shape. The air is held inside by three bladders of Combitherm material, commonly used in the food-packing industry. The innermost layer, forming the inside wall of the module, is Nomex cloth, a fireproof material that also protects the bladder from scuffs and scratches.

talking to Lockheed Martin on its plans for a CTV capsule and is working with Boeing on a CTV prototype. Bigelow also is keeping an eye on the Dragon CTV that SpaceX is developing for launch on the Falcon-9.

After launching his space stations into orbit, Bigelow has his sights set on the Moon, where he intends to deploy a base (Figure 5.14) capable of housing up to 18 astronauts in inflatable – what else? – modules on the lunar surface. Bigelow's lunar plans will only be implemented once they've made LEO work first, which will mean establishing a robust infrastructure in Earth orbit and creating the economies of scale necessary to produce facilities in LEO. For example, Bigelow is sketching out architectures to place their structures at Lagrangian Point, L1, where the habitats would be used as depots for missions to the Moon. Other architectures call for the deployment of habitats even larger than the cavernous BA-330; since NASA is working on a super-heavy lifter, these habitats are on the drawing boards. Bigelow envisions expandable habitats offering 2,100 cubic meters of volume (the BA-2100 – nearly twice the capacity available on the ISS) while another plan sketches out use of a super-jumbo structure providing 3,240 cubic meters of volume (the pressurized volume of the ISS by comparison is 837 m^3).

As you can see in Figure 5.14, the Moon base would utilize three BA-330s, along with topped-off propulsion tanks and power units, which could be joined together and then migrated from either L1 or lunar orbit and flown to pre-selected lunar locations. The expandable systems would be configured to be independent of each other and self-sustaining. Once operational, astronauts would test systems required for ventures to the Martian surface and beyond. The transport for such a trip would be the BA-2100, which could be placed into orbit by NASA's planned heavy-lift launcher. Once in orbit, the massive structure would be inflated and outfitted with materials carried aboard additional launches.

130 The New Breed of Space Vehicles

Figure 5.14 Artist's rendering of Bigelow Aerospace's Moon base. Courtesy: Bigelow Aerospace

With plans to orbit a space station, establish a Moon base, and venture to Mars, Bigelow Aerospace seems to have taken over the reins of NASA, an institution that not so long ago gave us a shot at interplanetary exploration, but which is now mired in budget cuts, increasing criticism, and a lack of direction. Against the backdrop of canceled programs, mismanaged budgets, and a non-existent manned exploration strategy, the vision of Bigelow Aerospace is a heartening reminder that mankind's reach into space isn't necessarily determined by the whims of one government's decisions.

SpaceX

It should have come as no surprise when SpaceX was named one of the 50 most innovative companies in the world by MIT's *Technology Review*. After all, SpaceX was founded with the goal of revolutionizing rocket and space technology, and to achieve that goal has required a fair amount of innovation since the company was founded in 2002. In selecting the 50 most innovative companies, the editors of *Technology Review* looked for those that had demonstrated superiority at inventing technology and were using it to grow as a business, and also for deploying and scaling up their technologies. They also looked at the likelihood that they will succeed. SpaceX was a slam-dunk on all counts. Not only will their Dragon spacecraft be capable of carrying the same number of astronauts as the retired Shuttle at one-tenth of the cost, but the vehicle will also be much safer due to technological improvements such as a launch escape system, automatic stability on re-entry, and having a more robust heat shield. In addition to its capacity for

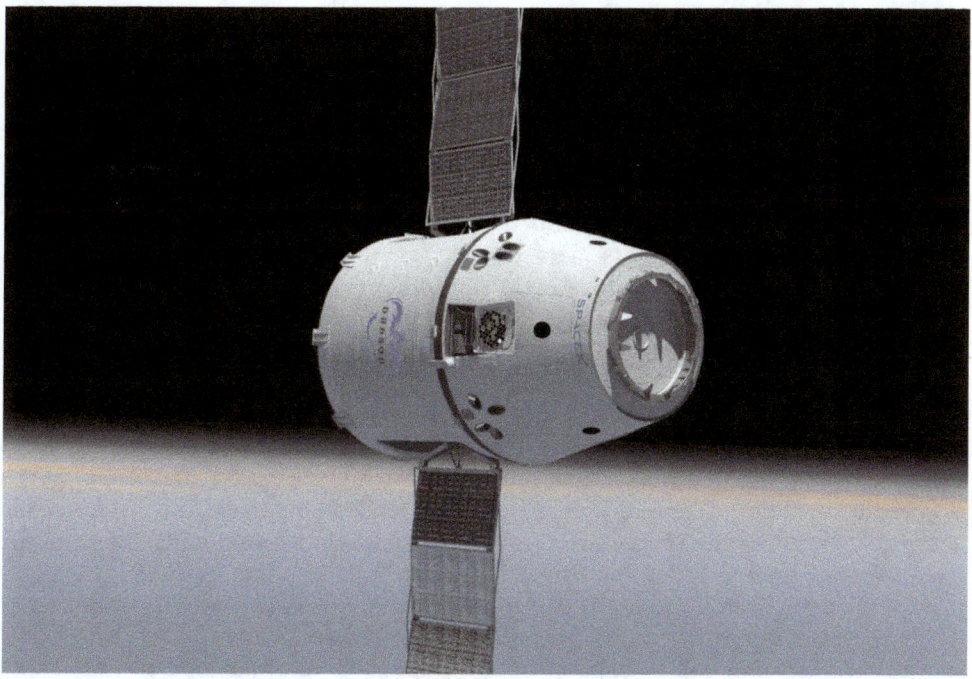

Figure 5.15 The Dragon capsule in orbit. Courtesy: SpaceX

innovation, SpaceX also has a solid business model, as evidenced by the company having grown to over 1,250 employees since it was founded (the company has been running a profit from 2007 to 2010).

On December 8th, 2010, Elon Musk's rocket company became the first commercial company in history to re-enter a spacecraft from Earth orbit when SpaceX launched its Dragon spacecraft into orbit atop a Falcon-9 rocket from Cape Canaveral Air Force Station in Florida. After orbiting Earth twice, the Dragon (Figure 5.15) re-entered Earth's atmosphere and landed less than a mile from the landing zone in the Pacific Ocean, thus marking a feat previously performed by only six nations or government agencies: the US, Russia, China, Japan, India, and the European Space Agency (ESA).

The flight, which was the first under the Commercial Orbital Transportation Services (COTS) program, marked the second flight of the Falcon-9, propelled by nine Merlin engines, which generate 1,000,000 lb of thrust in a vacuum. First-stage separation occurred shortly after three minutes into flight, after which a single Merlin took over to boost the second stage to orbit. After stage separation, the nose cap at the front of the Dragon was jettisoned and the second stage fired for another four-and-a-half minutes. On achieving orbital velocity, the Dragon separated from the second stage to begin independent flight, at which point the second stage restarted to carry the second stage to an altitude of 11,000 km – this wasn't necessary for ISS missions, but served to demonstrate the spacecraft's potential to customers who require payloads delivered to Geosynchronous Transfer Orbit (GTO).

After completing its designated orbits, the Dragon fired its Draco thrusters to begin the six-minute de-orbit burn. During re-entry, temperatures reached more than 1,650°C, but the Dragon's Phenolic Impregnated Carbon Ablator (PICA)-X^5 heat shield – the most advanced heat shield ever to fly – protected the spacecraft. At about 3,300 m, Dragon's three main parachutes, each 35 m in diameter (even if Dragon were to lose one of its main parachutes, the remaining chutes would still ensure a safe landing), deployed to slow the capsule's descent, ensuring a comfortable ride that will be required for manned flights.

Dragon

SpaceX started development of the Dragon with internal funding in 2005. The capsule (Figure 5.16), a pressurized, reusable system designed to carry either cargo or people, contains a reusable cargo segment called the spacecraft, a nosecone designed to protect the spacecraft's forward docking hatch and collar, and an aft unrecoverable section called the trunk. The spacecraft also features the PICA-X to protect the crew from the heat of re-entry. Two solar arrays are affixed to the trunk, where the life-support system and other components are located.

The cargo version will be capable of carrying 6,000 kg of payload to ISS or any point in LEO and transporting up to 3,000 kg of material back to Earth. To maneuver itself in orbit, the Dragon will use a reaction control system of 18 Draco thrusters. The trunk will house two solar arrays, providing an average of 1,500 watts of power. Before berthing with the ISS, the nosecone section will open to reveal a Common Berthing Mechanism (CBM) that allows Dragon to connect with the US Harmony module. However, the CBM does not enable docking with any of the four Russian modules, where Soyuz, Progress, and ATV dock. Docking with Russian components requires using an Androgynous Peripheral Attach System (APAS) that allows docking at multiple locations on the ISS. Upon completing a mission, the spacecraft section will disconnect with the unrecoverable trunk section to prepare for re-entry. After separation, the Dragon spacecraft will re-enter the atmosphere, with the assistance of an ablative heat shield. Externally, the passenger version of Dragon appears similar to the cargo version, since both have windows, but, in the spacecraft section, there are seven seats and a life-support system.

---

[5] NASA made its facilities available to SpaceX, as the company designed, developed, and qualified the PICA-X shield in less than four years at a fraction of the cost the agency had budgeted for the effort. The PICA-X is designed to be used hundreds of times for Earth orbit re-entry with only minor degradation and can withstand the much higher heat of a Mars re-entry.

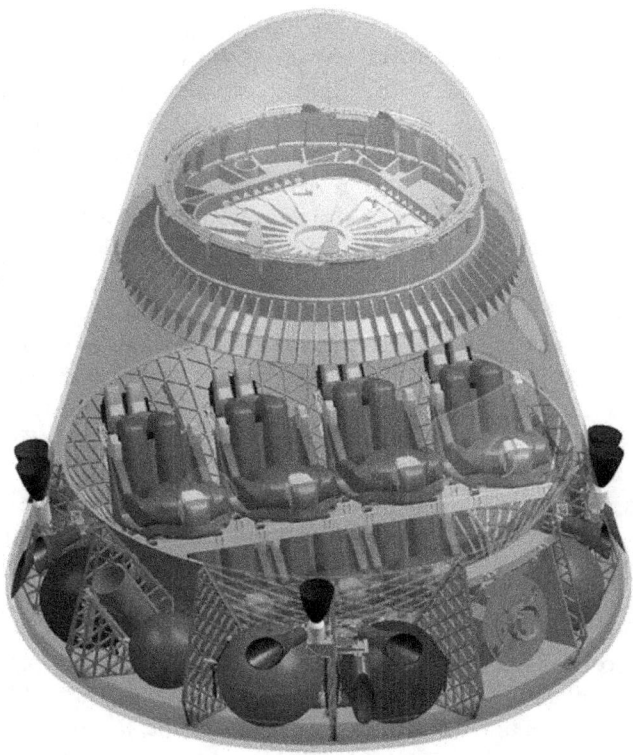

Figure 5.16 Cutaway of the Dragon capsule showing seating configuration. Courtesy: SpaceX

BOEING CST-100

Fancy a ride to the ISS but don't want to be launched into orbit on board the dated Soyuz? Well, Boeing is now in the commercial spaceflight business and will be flying astronauts for hire and rich spaceflight participants aboard its Crew Space Transportation-100 (CST-100) spacecraft sometime around 2015. Boeing says it has agreed to market the spaceflight participant flights through Space Adventures, which has already flown seven private individuals to the ISS aboard the Soyuz. Extra seats on the CST-100 (Figure 5.17) will be available to companies and non-governmental organizations that may employ commercial astronauts to fly their payloads.

The versatile CST-100, which can be deployed from different kinds of launch vehicles, may also be used on missions to LEO. The vehicle may also be used to ferry sovereign customers to Bigelow's inflatable orbital space habitat. Unlike commercial spaceflight companies, Boeing is hardly starting from scratch, although it recognizes that it needs to do things in an innovative way to make commercial spaceflight work. When operational, the CST-100 capsule, which will be able to fly up to 10 missions,

134 The New Breed of Space Vehicles

will be capable of remaining docked at the ISS for up to seven months, serving as a lifeboat in the same way as the Soyuz has done. On its return to Earth, the vehicle follows a similar return profile to the Apollo capsule, except the CST-100 touches down on ground. As you can see in Figure 5.17, the CST-100's blunt-nose capsule design is similar to the Apollo capsule, which isn't surprising given that Boeing is leaning heavily on Boeing heritage from Apollo.

While, outwardly, it may look like the Apollo capsule, it will be larger than its famous cousin, although Boeing hasn't released details of the vehicle's dimensions. It's not that Boeing is secretive; its just that it's been busy completing other tests. You see, under the CC-Dev agreement with NASA, Boeing had to achieve a series of technological and development milestones within a specified timeframe to be eligible for more funding – it received $18 million in the first round. The last of the CCDev-1 milestones was a series of hot-fire tests of the Bantam demonstration engine that may be used as part of an innovative "pusher" launch abort system (Panel 5.2). A pusher launch abort system "pushes" or propels a spacecraft towards safety if a launch abort is needed and, if unused for an abort, the propellant can be used for other portions of the mission. Boeing's tests were successful, meaning they were not only eligible for more funding, but they validated the pusher launch abort system; with this system, the CST-100 could use the abort fuel to re-boost the ISS orbit, which is an added benefit to NASA. NASA was happy with Boeing's demonstration, awarding the company an extra $92.3 million in the second round of CCDev funding. The money allows Boeing to continue development of the CST-100; its next goals include a preliminary design review (PDR) in the fall of 2011 and a critical design review (CDR), the last major step before actual construction of the vehicle, about a year later. That would put Boeing on a path to conduct a pad abort test of the CST-100's escape system in 2013, followed by two un-crewed test flights in 2014 and finally a flight with two test pilots around 2015.

Russian Soyuz and Chinese Shenzhou spacecraft use rockets attached to the top of the crew capsules to drag the crew out of harm's way; the same type of launch abort system was first employed in NASA's Mercury program and became familiar

---

Panel 5.2. Escape systems

Unlike fighter pilots, who have the option of using an ejection seat in the event of an emergency, for every astronaut to be thrown clear to parachute height on an ejection seat, they all have to be sitting near an escape hatch, and fitting too many hatches weakens a spacecraft's structure. The Shuttle used ejection seats during its first four test flights but the seats were removed after the Shuttle was declared operational (the seats were only of use at speeds of up to Mach 3, and the Shuttle reached Mach 18). After the *Challenger* disaster, the Shuttle got a (very!) limited bail-out capability allowing the crew to jump from a hatch with parachutes if the vehicle was gliding at an altitude of less than 6 km.

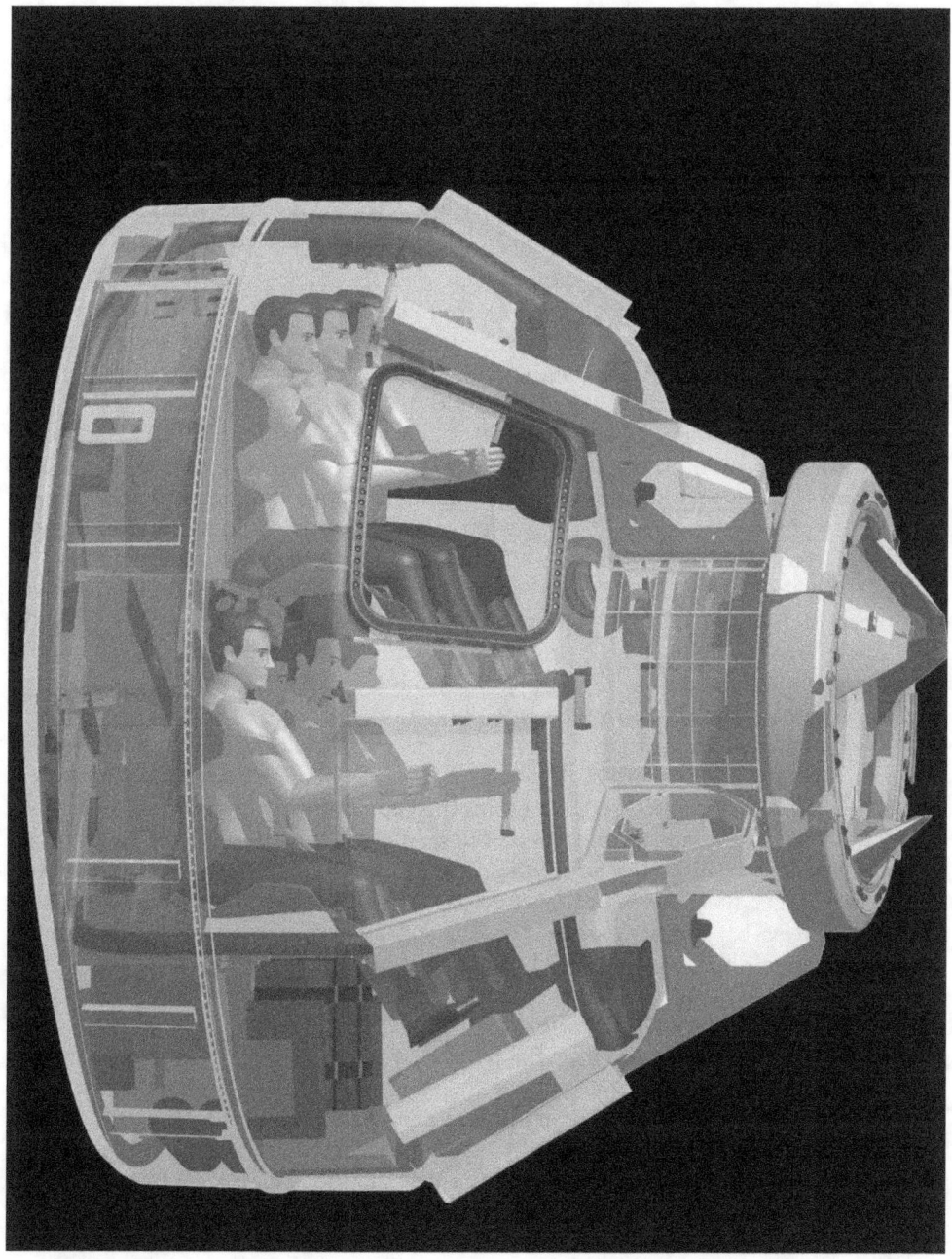

Figure 5.17 Cutaway of Boeing's CST-100 capsule showing seating configuration. Courtesy: Boeing

in the Apollo program. Apollo used a 12-m-long tower fixed to the top of the crew capsule. About a third of the way up the tower, four solid-rocket booster nozzles pointed downwards and away from the capsule. In an emergency, explosive bolts would release the capsule from the launch rocket, and the solid-fuel escape rockets would shoot the capsule to a parachute-deploying altitude. The problem with these designs is that, if the escape system isn't used during lift-off, it's jettisoned once the air thins out but, until then, it adds fuel-eating weight, wrecks the capsule's aerodynamics, and dumps some perfectly good thrusters into the ocean. If the escape system *is* used, a partial vacuum develops between the capsule and the launch rocket below it as the capsule begins to pull away, creating extra drag that makes it harder for the capsule to fly to safety. The alternative is the "pusher" design that would propel the capsule from below. A service module beneath the capsule would carry thrusters, powered by oxygen and liquid fuel that would be used to move the capsule in orbit. The advantages of a "pusher" system include cleaner aerodynamics and putting escape thrusters on the bottom of the capsule pressurizes the gap between the capsule and its launch rocket when they separate, eliminating the suction-cup effect created between the two in traditional designs. The disadvantage of the pusher system is that it leaves a lot of fuel beneath the crew for their trip, which adds to the risks in orbit.

Now that the Shuttle fleet has retired, many experts argue it will be five to seven years before the US has its own manned space capability. However, with the development of the CST-100, Boeing could shave two or three years off astronauts' down time without something American to fly. Even if the CST-100 is delayed, you now have private rocket and spacecraft builders racing to the finish line with new hardware: two capsules, a space plane, and a gumdrop spaceship to taxi astronauts to and from the ISS or other destinations in LEO. Boeing's CST-100 spacecraft is obviously the Cadillac of the entries in this commercial space race – assuming adequate NASA funding continues. Of course, there is still the issue of which rocket to use. There has been talk of using the Atlas 5 or the Delta IV but the most likely candidate may be the Liberty, a rocket offered by ATK Space Launch Systems. The rocket (Figure 5.18) is arguably the world's most reliable rocket: a US–European vehicle that is an upgraded version of the Space Shuttle's solid booster rocket, which has flown perfectly 216 times, attached to France's Ariane 5 rocket as a second stage, which has flown 41 times without a problem.

The new Liberty launch vehicle, which draws on parts of NASA's cancelled Ares I rocket (which, in turn, drew on parts of the retired Shuttle) and Europe's Ariane 5 rocket, will use existing infrastructure at Kennedy Space Center, such as the Mobile Launcher shown here in Figure 5.19. Developed by ATK, a longtime NASA contractor, the five-segment (one segment longer than the solid-rocket boosters used on the Shuttle) solid rocket's original purpose was to serve as the first stage for NASA's Ares I rocket for launching astronauts into space on Orion capsules. However, this fell through when President Barack Obama cancelled NASA's Constellation program aimed at returning astronauts to the Moon. Following the cancellation of Constellation, ATK teamed up with the European aerospace company Astrium to repurpose the Ares I rocket's first stage into a completely new, commercial launch vehicle: the Liberty.

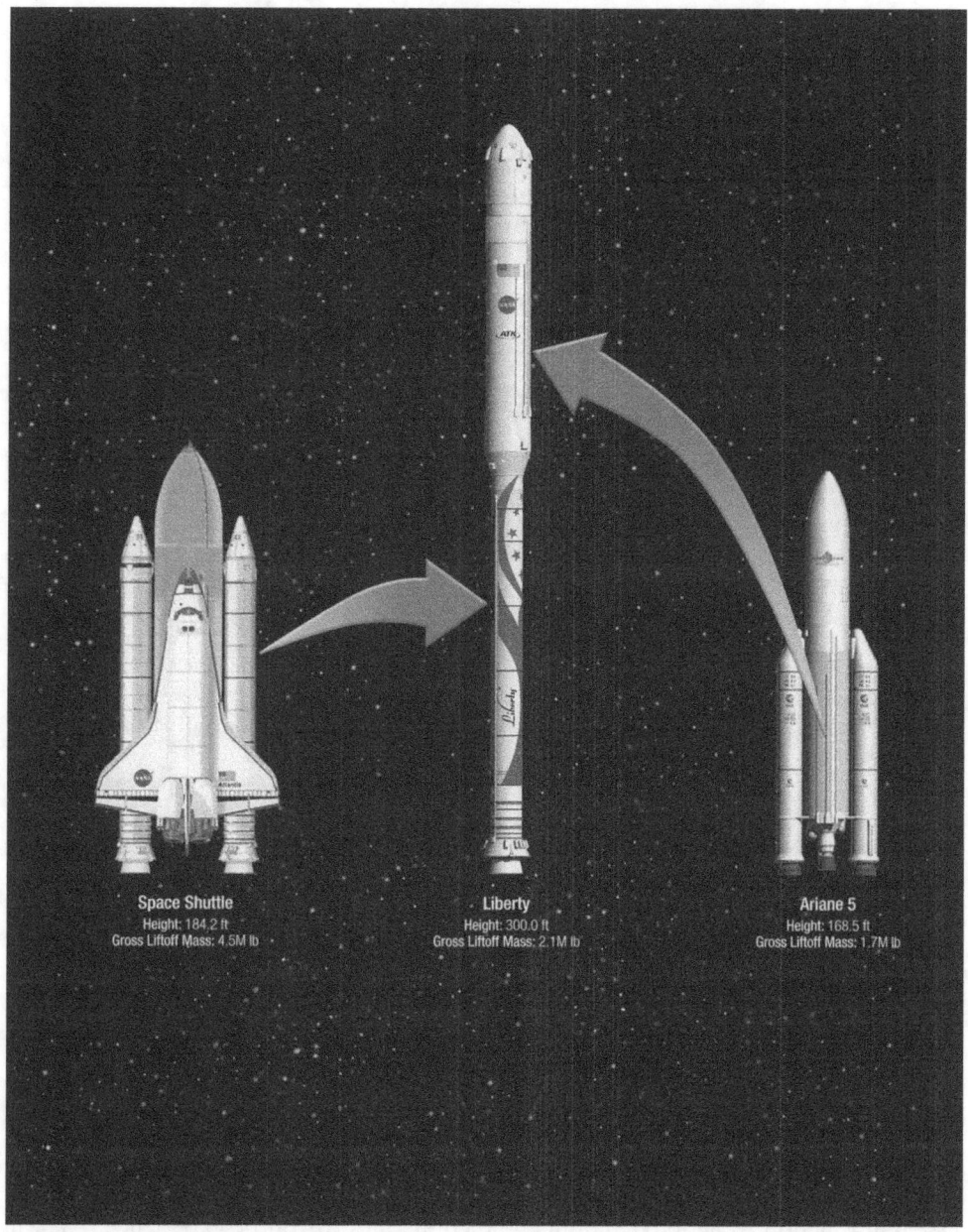

Figure 5.18 The Liberty launch vehicle combines the proven systems from NASA's Space Shuttle fleet and Europe's Ariane 5 expendable rocket. Courtesy: NASA

Figure 5.19 The new Liberty launch vehicle, which draws on parts of NASA's cancelled Ares I rocket and Europe's Ariane 5 rocket, will use existing infrastructure at Kennedy Space Center, such as the Mobile Launcher shown here in this illustration. Courtesy: NASA

ATK and Astrium officials say the Liberty rocket could be ready for a first test flight by 2013 and begin operational launches by 2015 from NASA's Kennedy Space Center (KSC) in Florida. Astrium builds the Ariane 5 rockets that have served as Europe's primary workhorse for launching satellites and spacecraft into orbit. The company would provide the second stage for the new rocket that is based on the Ariane 5 vehicle, officials said. Since both stages were designed for human-rating (a time-consuming process that the other rockets haven't been subject to) since inception, there wouldn't be a problem launching the CST-100. Infrastructure won't be a problem either; a mobile launch platform and launch pad modifications at KSC were built for the Ares I rocket to support its one and only test flight (called the Ares I–X) in 2009.

The fact that Boeing, with its vast manned spaceflight experience stretching over decades, is leading a CCDev team gives enormous credibility to the commercial spaceflight venture. One of the oft-heard criticisms is that NASA is turning over an essential function to neophyte commercial companies with little experience that will not be able to fly safely or have deep-enough pockets to stay in the game if economic conditions get tough. It's a harder case to make with Boeing involved. The four pioneering commercial space companies developing the capability to take crews to LEO may inaugurate a new era – a transformative one, fuelled by entrepreneurship and supported by markets ranging from missions and space tourism to scientific research.

Section III
The Missions

6

Suborbital Missions

Cape Canaveral, Florida, February 24th, 2011. After a delay of nearly four months for fuel-tank repairs, *Discovery*, the world's most traveled spaceship, thundered into orbit for the final time, heading towards the International Space Station (ISS) on a journey that marked the beginning of the end of the Shuttle era. *Discovery*, the oldest of NASA's three surviving Shuttles, and the first to be decommissioned, was embarking upon its 39th mission and the 133rd mission of the Shuttle program. There were several tense minutes before lift-off when an Air Force computer problem threatened to halt everything, but the issue was resolved and *Discovery* blasted off just three minutes late. Not surprisingly, emotions ran high as *Discovery* (Figure 6.1) rocketed off its seaside pad one last time, the resonant thunder of its main engines and the crackle of its solid-rocket boosters rolling out across the Florida countryside into a late-afternoon clear blue sky, before arcing out over the Atlantic.

With the retirement of the Shuttle fleet rapidly approaching, it wasn't surprising that *Discovery*'s last flight drew a sizable crowd, with an estimated 40,000 guests gathered at Kennedy Space Center (KSC) to witness history in the making. Roads leading to the launch site were jammed with cars parked two and three deep and recreational vehicles snagged prime viewing spots along the Banana River well before sun-up. By the time the mission had ended, *Discovery* had racked up nearly 150 million miles on the clock, spent 363 days in space, and circled Earth 5,800 times. No other spacecraft has been launched so many times, nor has any Shuttle racked up so many achievements; more crew members have flown in *Discovery* than any other space vehicle, including the first female to pilot a US spacecraft and the oldest person to fly in space. Sadly, it was also the first Shuttle to retire.

Four days later, in the bar of the Holiday Inn near the University of Central Florida (UCF), I'm talking with Mindy and Joe, two fellow Astronaut for Hire (A4H) colleagues. We're discussing the next step in manned spaceflight, which happens to be the subject of the meeting that has brought us to the UCF: the second annual Next-Generation Suborbital Researchers Conference (NSRC). The main topic of conversation at the conference is the news that the Southwest Research Institute (SwRI) has budgeted US$1.3 million for a four-year suborbital science

Figure 6.1 Space Shuttle *Discovery* launches on its 39th and final mission. Courtesy: NASA

program, a portion of which will be used to book passenger seats on spacecraft for Alan Stern and his SwRI colleagues, Dan Durda and Cathy Olkin. If all goes as planned, the three researchers will be flying into space as fully fledged astronauts by mid 2013.

The SwRI deal is a slice of commercial space history, since the agreement is the first of its kind, which isn't that surprising, since none of the leading suborbital companies is planning on flying its vehicles much before mid 2013 at the earliest. Some space scientists remain dubious about the usefulness of commercial suborbital research flights, but those scientists weren't around at NSRC, where the prevailing mood was raw enthusiasm. And, of all the attendees, no one embodied the upbeat feeling better than Stern. As former Associate Administrator for NASA's Science Mission Directorate and the principal investigator for the agency's New Horizons mission to Pluto, Stern has a reputation for making big things happen; put simply, NSRC wouldn't have happened without Alan and SwRI. When interviewed about the promise of commercial spaceflight, he often compares the industry to the early days of commercial aviation and goes on to describe the day when there will no longer be one centrally planned space program in which only NASA has the keys to

Figure 6.2 Zero-G Corporation's specially modified Boeing 727, G-Force One. Weightlessness is achieved by conducting aerobatic maneuvers known as parabolas. Before starting a parabola, G-Force One flies level to the horizon at an altitude of 7,200 m. The pilot begins to pull up, gradually increasing the angle of the aircraft to about 45° to the horizon, reaching an altitude of 10,000 m. During this pull-up, passengers feel the pull of 1.8 G. The plane is then "pushed over" to create the zero-gravity segment of the parabola. For the next 20–30 seconds, everything in the plane is weightless. Next, a gentle pull-out is started, which allows the flyers to stabilize on the aircraft floor. This maneuver is repeated 12–15 times, each taking about 16 km of airspace to perform. Courtesy: Zero-G Corporation

space. Instead, space travel will be offered by companies such as Virgin Galactic and XCOR, who both sent representatives to NSRC. These entrepreneur spaceflight companies expect one of the most lucrative applications for suborbital spaceflight to be space tourism, but they have also begun courting research institutions and government agencies who want to run scientific experiments in suborbital space – companies like the SwRI, which bought eight seats, and options for nine more, on suborbital flights, split between Virgin Galactic's SpaceShipTwo and XCOR's Lynx.

None of my A4H colleagues needed convincing about the potential of suborbital research missions, but the enthusiasm of researchers is one thing – getting the institutions they work for to pay for tickets is something else. To sell those passenger seats, Virgin Galactic and XCOR will have to convince decision-makers that their vehicles offer significant advantages over existing ways to access microgravity. After all, there are cheaper alternatives; for only a few tens of thousands of dollars per flight, researchers can book modified 727s (Figure 6.2) that fly a series of parabolic

144   Suborbital Missions

trajectories, providing microgravity in 24–30-second bursts. And then there are NASA's sounding rockets.

Sounding rockets (Panel 6.1), like the Black Brant depicted in Figure 6.3, carry scientific instruments into space along parabolic trajectories. Although the time in space is brief (typically 5–20 minutes), for a well-placed scientific experiment launched into a geophysical phenomena of interest, the short time and low vehicle speeds are more than adequate to carry out a successful scientific experiment. Another reason scientists like to use sounding rockets is to access certain regions of space such as the ionosphere and mesosphere that are too low to be sampled by satellites. Because the science payload doesn't go into orbit, sounding rockets don't need expensive boosters or extended telemetry and tracking coverage, which means mission costs are much lower than those required for orbital missions. Other savings

---

Panel 6.1. Anatomy of a sounding rocket

A sounding rocket basically comprises solid-fueled rocket motor(s) and a payload. The Black Brant, one of the most popular sounding rockets, has a two-stage solid-fueled rocket; the first stage is a Terrier booster and the second stage is a Black Brant motor. The payload obviously depends on the mission, but there are some components that most payloads have in common. To begin with, there is the Nose Cone/Ogive Recovery System Assembly (ORSA), which contains the parachute and the Attitude Control System (ACS) pitch and yaw jets which control the payload orientation. A sounding rocket also has to have a Boost Guidance Control (BGC) to reduce impact dispersion and acquire accurate trajectories. Using the BGC, the rocket is able to launch to higher altitudes and in higher winds thanks to gyros that detect pitch and yaw angles and pneumatic servos that turn pairs of canards to adjust the trajectory during the boost phase of the flight. The telemetry section is the nervous system of the rocket payload, transmitting data to the ground station, including housekeeping data and science data. It can also receive uplink commands that steer the rocket. A High Velocity Separation System (HVSS) connects the rest of the payload to the motors by a v-band, which is held together with four screws; at the appropriate time, these are cut by explosives and the v-band released. This allows four springs to expand and push the payload from the rocket. Finally, there is the igniter housing, which contains the yoyo-despin system and thrust termination system. After the rocket has burned out, it's spinning at about four cycles per second. The yoyo-despin releases a weight attached to a long wire coiled about the rocket, thus carrying away angular momentum. Any residual spin is adjusted for by the ACS roll jets. The thrust termination system, consisting of explosives affixed to the motor, is used if the impact dispersion is greater than the safety limits.

Figure 6.3 Black Brant sounding rocket. Courtesy: NASA

are achieved thanks to the payloads being built in one central location and the high degree of commonality and heritage of rockets, payloads, and subsystems that are flown repeatedly; in many cases, only the experiment, which is provided by the scientist, is changed. Then there's the quick turnaround; a typical payload can be developed in as little as three months – a timeframe enabling scientists to react quickly to new phenomena and to incorporate the latest, most up-to-date technology in their experiments.

So, sounding rockets are robust and versatile. They also provide critical scientific, technical, and educational contributions, can fly relatively large payloads, provide several minutes of microgravity, and can gather in-situ data in specific geophysical targets such as the aurora, noctilucent clouds, and thunderstorms. Cost-effective? Not so much; a flight on a Black Brant sounding rocket can cost over $2 million, which is about an order of magnitude more than a seat on a SS2 flight. Also, sounding rocket flights are relatively infrequent, flying only a few times a year, while some suborbital vehicles are planning to fly multiple times a day.

Furthermore, unfortunately, sounding rockets are unmanned and 24 seconds of microgravity doesn't compare with the four to five minutes provided by suborbital providers. But, is the risk of having a human ride along with an experiment worth it? Well, having a human on board does increase the risk and adds to the complexity (just the weight of seats and life-support exacts a huge performance penalty), but a human can easily react to an investigator's data collection in real time and make changes in the experiment to get better results.

Some argue the real benefit to science may not be realized with the first wave of vehicles, but in the second or third generation of vehicles, which will probably reach LEO. Orbital capability married with an ability to routinely and frequently access longer periods of weightlessness could translate into imaginative science. Right now, as the NSRC attendees listened to spaceflight operators forecasting revenue flights some time in 2013 or beyond, the main interest seemed to be a yearning to simply go into space, science or no science. It was a motivation that Stern alluded to in one of his NSRC sessions when he told the audience that suborbital spaceflight shouldn't only be measured in grants and peer-reviewed papers. "I don't think most scientists appreciate very well how motivational human spaceflight can be," Stern said. "We can contribute to the future by giving birth to this new industry and the opportunities it brings." It was a message that resonated strongly at the NSRC; at the coffee stand between sessions, discussions about suborbital science often touched on the subject of inspiring the next generation of researchers – after the current generation had flown of course!

> "You spark this industry with tourists, but I predict in the next decade the research market is going to be bigger than the tourist market."
>
> Alan Stern, SwRI

Alan's probably right. At least I hope he is, because I don't have $200,000 to plunk down for a ride on board SS2; like Alan and Dan, I, along with several NSRC attendees, have gone through the traditional route and made it as far into the process as you can without being selected. Now we're going to fly anyway. Whether the tourist market is big enough to sustain private-sector spaceflight is open to speculation, but if there's one thing that all the major players in the gestating suborbital industry agree on, it's that research flights could make the difference.

Aabar Investments certainly seems to think so. In 2009, the Middle East investment fund that recently bet big on Mercedes-Benz (it bought 9.1% of the German auto maker in March 2009) paid about $280 million to buy nearly a third of Virgin Galactic. Aabar Investments' buy-in gave Sir Richard Branson's space tourism venture a big financial kick-start at a time when many funding sources had dried up because of the global recession. It also gave the wealthy Persian Gulf sheikdom of Abu Dhabi a chance to build its own spaceflight industry as it broadens its economy beyond the oil sector. In exchange for its 32% stake in Virgin Galactic's holding company, the state-controlled fund will acquire "exclusive regional rights" to eventually launch Virgin Galactic scientific research spaceflights from the United Arab Emirates (UAE) capital.

Sending experiments into space on a for-profit spacecraft will be very different from elbowing for room on a federal rocket. Cheaper costs aside, scientists at universities and private research institutions could face some new legal questions on the subject of intellectual property and technology licensing agreements. The 15 member nations of the ISS have patent agreements covering various modules on the ISS and how to handle discoveries made in those modules. For example, any work done inside the Japanese Experiment Module KIBO is considered Japanese property. Similarly, commercial spacecraft companies may craft contracts that ensure a royalty-free license to any new discoveries made on their vehicles, meaning a scientist would own any invention that comes from a suborbital eureka moment. Alternatively, and more likely, the company that made such an invention possible may want to get a piece of the action, just like a privately owned laboratory would want a license to use an invention made at the laboratory.

The research proposed for these vehicles is diverse. For example, Josh Colwell, a planetary scientist at the University of Central Florida, is interested in studying the dynamics of ring particles in microgravity and comparing those to models of planetary and ring formation. Another researcher interested in the microgravity environment suborbital vehicles will provide is Steven Collicott of Purdue University, who studies fluid dynamics in low-gravity environments. Colwell and Collicott are scientists who have already grabbed the suborbital bull by the horns and have struck deals with vehicle providers: in this case, Blue Origin (incidentally, Alan Stern happens to be Blue Origin's advisor for REM applications). Blue Origin began soliciting investigator experiments to be flown on a "no exchange of funds" basis in 2009. In September 2010, after spending 18 months studying the research and education market, the famously quiet rocket company (one of the reasons given for the company's low profile is that it keeps their marketing and public relations staff small!) lifted the veil of secrecy and announced it had selected three experiments to fly on Phase 1 of the development program for its New Shepard suborbital vehicle. Joining Colwell and Collicott is John Pojman, a chemistry professor at Louisiana State University (LSU), who is looking forward to quick progress on his research into fluid dynamics – specifically, how fluids of different densities mix under various conditions. Years ago, Pojman was in line to have one of his experiments flown on the ISS. His LSU team had passed the scientific review for an experiment that involved injecting a squirt of Russian honey into water recovered from the station's urine collection system. To work out the details of the experiment took 900 e-mails and numerous teleconferences over the course of six months. Then, after *Columbia* disintegrated in 2003, his team was de-flighted. Hardly surprising, then, that Pojman welcomes the era of suborbital science flights!

Not surprisingly, there is great interest in using suborbital vehicles in life sciences, particularly for studying human reactions to microgravity. One thing that sets these vehicles apart from microgravity flights or orbital spaceflight missions is the potentially large, diverse group of people who will be flying on these spacecraft as potential test subjects. For the last 50 years, space physiologists have been studying very healthy astronauts, most of whom are between the ages of 35 and 50, which is a very narrow population. With the advent of suborbital missions, space life sciences will be extended to a much larger population group, including some with chronic health problems, which may lead to some interesting results. But, will the well-heeled space explorers, who'll be paying up to $200,000 for their trip, be willing to volunteer as test subjects? Perhaps. Perhaps not. We'll have to wait and see.

While Alan Stern's Virgin Galactic flights and Blue Origin's agreement to fly experiments sound promising, getting more experiments on more vehicles will require more than just outreach to researchers and vehicle developers; it will also require the support of NASA, a major source of funding for such experiments. In the past few years, the agency has given mixed messages about the level of interest in funding research on commercial suborbital vehicles, but that attitude seems to be changing. In February 2010, NASA deputy administrator Lori Garver announced the creation of the Commercial Reusable Suborbital Research (CRuSR, pronounced "cruiser") program to foster development of suborbital research efforts with commercial providers. The $15 million allocated for CRuSR will help researchers procure rides into space for their experiments and, ultimately, themselves. Some observers interpreted the announcement that NASA would be purchasing rides for researchers in the near term without oversight, which prompted the agency to clarify the issue by saying:

> "NASA would like to clear up some misunderstandings related to the CRuSR Program. The CRuSR Program plans to purchase flight services for untended payloads supporting science, technology and education. NASA will NOT be paying for humans to fly on the commercial vehicles until NASA determines the vehicles meet NASA standards for human safety. NASA will treat these vehicles as the high-speed experimental aircraft they are when assessing safety. Dryden, which has 60 years experience flying high-speed experimental aircraft, including the X-1 and X-15, will lead the safety assessment. This NASA process will only apply to NASA-funded human passengers."

Basically, it's likely that NASA won't fly researchers on a given system until they've flown NASA hardware first, but that won't happen until researchers start winning funding for suborbital science proposals.

FLYING A SUBORBITAL PAYLOAD

So, how does one go about flying an experiment on a suborbital vehicle? After all, you can't buy a ticket from a company like a tourist can from Virgin Galactic – not yet, anyway. The whole process will begin with an announcement of opportunity being made by a space agency or research institution. After coming up with an idea and filling in lots of forms, the application will be sent to the agency for review (Table 6.1). Once accepted, the scientists, led by a Principal Investigator (PI), will embark upon a research design for the flight, which could take up to six months or longer. Some time during the research design period, the PI will receive an experiment form from the operator, requesting that the PI provide them with the particulars of all the scientists involved in the experiment, experiment objectives, experiment description, and a technical description of the experiment set-up. This will require the PI to provide a description of each system, an explanation of each experiment rack (including designation, function, mass, dimensions, etc.), schematics of the experiment, details of electrical circuits, a list of all products, photographs of the experiment set-up, and, finally, the team's approach for designing, building, and testing the experiment. Other items on the experiment form will include details such

Table 6.1. Suggested timeline for a suborbital flight experiment.

	L -- 36	L -- 30	L -- 24	L -- 18	L -- 12	L -- 6	L -- 3	Launch
Submit proposal	▲							
Proposal review	←——→							
Proposal acceptance			▲					
Research design			←—————————→					
PI receives experiment form				▲				
PI confirms participation					▲			
Medical exam				←→				
Operator receives experiment form						▲		
PI and scientists visit operator							▲	
Execute changes to research							←→	
Design frozen								▲
Medical docs submitted to operator								▲
Test of payload								▲
Safety review meeting							←→	
Changes if necessary							←→	
Liability form submitted to operator							←→	
Preflight training								←→

as electrical consumption, a mechanical resistance analysis, in-flight procedures (a list of major tasks the scientists will perform during the flight), certification for use of human subjects, a liability waiver, an experiment hazard evaluation, and a hazard list. It's a lot of paperwork, but it doesn't stop with the experiment form because, two months before the flight, the equipment data package form arrives. This requires more specific information including a structural load analysis, proof of mechanical resistance of each structure, detailed test procedures, the particulars of the data acquisition system, and test operation limits and restriction.

Once that round of paperwork is out of the way, the team can look forward to the Safety Visit. The Safety Visit is the final review before the flight and allows the operator to inspect test equipment, check any modifications to the payload, and approve or disapprove the test equipment. Around this time is when the team has to submit their medical certificates to the operator. After the Safety Visit, the team head back home and work on any modifications to the payload. Then, a week before the flight, they pack their bags and travel to the spaceport to begin their preflight preparations (Table 6.2). One of the first items on the schedule will be more

Table 6.2. Suggested events leading up to suborbital science launch.

Time	Event
L − 7 days	Scientists and team members arrive at spaceport; review by Principal Investigator
L − 6 days	Commence experiment preparation at Spaceport science staging facility
L − 5 days	Configure experiment into spacecraft
L − 4 days	Commence preflight training Day #1 of 4. AM: Academic instruction on high-altitude indoctrination and high acceleration; PM: Slow and rapid decompression in high-altitude chamber followed by emergency egress procedure training
L − 3 days	Preflight training Day #2 of 4. AM: Flight G-profile in centrifuge followed by task acquisition exercise review; PM: Zero-G preflight familiarization followed by pressure suit acquaintance and testing (donning and doffing)
L − 2 days	Preflight training Day #3 of 4. AM: Survival brief
L − 1 day	Preflight training Day #4 of 4. AM: PM: Flight safety briefing followed by presentation of each experiment by respective PIs
L − 1 hr	Scientists meet at Spaceport for final preflight debrief; optional anti-emetic medication given; scientists conduct final check of payload
L − 30 min	Scientists and crew board spacecraft; experiments are switched off
L − 20 min	Hatches are closed; passengers are requested to be seated and spacecraft begins taxiing; spacecraft electrical panel switched off
L − 10 min	Spacecraft electrical panel switched on
Launch	Spacecraft takes off
L + 15 min	Passengers leave their seats and switch on experiments
L + 20 min	Passengers switch off experiments and adopt landing configuration; electrical panel switched off
L + 35 min	Landing and debrief
L + 4 hr	Modifications and preparation of experiments for following day

paperwork; the operator will verify that all passengers have their medical certificates in order, that liability and waiver forms, if required, are signed, and all necessary modifications have been implemented. The operator will give the science team an overview of the week's activities and training. Once all the paperwork is out of the way, the scientists will begin positioning their payload in the spacecraft, assisted by the operator's check-out team, who will help the scientists with attachment interfaces and electrical input and output requirements. The loading, bolting, and electrical connecting will take about two days, after which the team will commence their preflight training, which will follow a schedule similar to that described in Chapter 3.

LAUNCH DAY

Launch day begins with breakfast served in the spaceport restaurant. While some may have an image of spacefarers enjoying a breakfast of steak, eggs, and coffee, the reality will be somewhat different because these astronauts for hire will probably be too damn nervous to eat anything more than half a slice of toast! However, in the early stages of commercial spaceflight, breakfast will be a mandatory photo opportunity for the media people, so the soon-to-be astronauts will probably fake a care-free smile and pretend to eat. After the meal, the crew performs one final pre-launch briefing before reviewing the launch countdown status and weather forecast. Next, the crew visits the flight surgeon for one final medical examination before visiting the bathroom for a final attempt to void their bladders before suiting up.

Unlike NASA astronauts, whose first challenge wasn't dealing with the acceleration stress following launch or adjusting to weightlessness, but donning the rather cumbersome launch and entry suit, our astronauts for hire only have to contend with slipping on a regular flight suit topped off with a helmet (Figure 6.4). However, there are a couple of suit technicians available to check the integrity of the suit and also the communication system. While the crew is suiting up, the vehicle is inspected in the same way as a regular aircraft is inspected. Using cameras and sensing devices, the team methodically checks every surface of the vehicle, looking for abnormal temperature readings or anything else that might indicate a problem. Once they finish their inspection, they report to the launch director.

Once suited up, the team wanders over to the vehicle and finds their seats. No close-out crew for this group of spacefarers – just a simple tug of the straps and hooking up to the communication system. The hatch is closed and a moment later the crew's ears pop as the capsule is pressurized. Meanwhile, back in Mission Control, the flight director thumbs through the countdown manual; it documents not only the launch procedures, but also the details of processing the spacecraft, the set of countdown instructions (Table 6.3) that sequentially configure the vehicle for launch, the instructions to be followed in the event of a scrubbed launch, and pre-planned contingency procedures and emergency instructions.

152 Suborbital Missions

Figure 6.4 Suborbital flight suit designed by Orbital Outfitters. Courtesy: Orbital Outfitters

Table 6.3. Launch countdown milestones.

Time		Event
T – HH	T – MM	
02	00	Crew eats breakfast and conducts media interviews
01	30	Crew suit-up begins assisted by suit technicians
01	00	Mission management team meeting
00	50	Avionics checkout and pilot walk-around of vehicle
00	45	Crew weather briefing
00	40	Crew ingress vehicle and commence pre-flight checks
00	35	Flight crew equipment stow
00	35	Communication activation
00	30	Crew communication checks
00	30	Science team ingress vehicle
00	25	Debris inspection
00	20	Ascent switch list
00	15	Pad clear of non-essential personnel
00	10	Final crew weather briefing
00	05	Hatch closure; vehicle pressurized
00	05	Area clear to launch
00	04	Launch Director launch status verification
00	02	Crew closes and locks their visors
00	00	**Launch**

Once the launch team reports to the flight director that there are no constraints to launch, the Launch Director gives his permission to proceed with the countdown. Meanwhile, inside the spacecraft, the crew performs radio checks and listens to the sequence of countdown milestones. As the clock shows the T – 10-minutes mark, the astronauts check their straps one more time, listens to the endless series of acronyms being checked off by the flight controllers, and watch their pilot follow the checklist. In just a few minutes, the vehicle will lift off from the pad in a thunderous cloud of fire and smoke. At two minutes before launch, the Mission Control Console (MCC) reminds the crew to close their visors – a command confirming launch is imminent. The crew oblige, snapping the bail into place to seal the faceplate. With adrenaline pumping and their heart rate close to the red-line in anticipation, the crew tries to slow their breathing and control the attack of butterflies. For a moment, the thought that these could be the final moments of their lives enters their head, but their attention is quickly drawn to the expectancy of the vehicle lighting up. After years of imagining riding a rocket into space, the rookie astronauts for hire are finally going to get their chance at wish fulfillment. Bodies tingling with anticipation, the rookies give each other a thumbs-up.

At one minute to launch, the primary flight displays (PFDs) continue scrolling data as the soon-to-be astronauts think about their families and all the training and preparation that brought them to this pivotal moment in their lives. The astronauts' families are a short distance away, standing on the roof of the spaceport, and are probably more scared than the astronauts. Mission Control announces 20 seconds to launch. At 15 seconds to go, the attitude indicators on the PFDs scroll to the correct launch attitude as the vehicle processes its final update.

Finally, the countdown reaches the endpoint. Ignition commands from the vehicle's flight computers commit the vehicle to flight as electrical energy is routed to the motor igniter, which initiates the burn of the rocket motor. A rumble shakes the vehicle as the main engine spews fire and roars to full power. On the PFDs, the computers scroll rapidly through engine checks and mark the mission time: zero. The clock is running! Inside the vehicle, the astronauts are pushed back into their seats by a force more than one-and-a-half times the force of gravity. The rocket engine is already consuming propellant voraciously; the crew are pushed further back into their seats as the gravitational forces ramp up to nearly 3 G. On the roof of the spaceport, the families watch awestruck as the vehicle blazes a trail of flame against the blue sky.

Less than 30 seconds after launch, shock waves form on the nose and the vehicle begins to vibrate. Meanwhile, the vehicle's control system continuously monitors trajectory and issues guidance commands to ensure the rocket nozzle's angle is correct. A few seconds later, the commander announces Mach 1 and the crew listen to the rushing sound of supersonic air being displaced by the vehicle. The commander gives the crew the signal that they can shut off the oxygen and open their visors. The vehicle is now more than two minutes into flight and flying at an altitude of more than 60 km. A few more seconds pass and the commander announces an altitude of 100 km. The rookies let out a cheer, since the altitude means they are now officially astronauts and have collectively achieved their life goal. Then they go to work conducting their assigned mission science tasks.

PORTFOLIO OF GAME-CHANGING MISSIONS

There is a bounty of science missions waiting to be tapped by the commercial suborbital spaceflight industry: everything from heliophysics to high-altitude meteorology and biophysics to bioscience. We'll discuss some potential missions here.

One popular suborbital science topic is atmospheric research. Atmospheric scientists are particularly interested in the impact of chemical transport at the turbopause (95 km) on neutral and ionized species higher up in the lower thermosphere. To date, despite deploying in-situ investigations, the middle atmosphere has proven to be dynamically complicated, so atmospheric scientists have suggested releasing chemicals from suborbital rockets to track winds and turbulence.

Another research topic that requires suborbital mid-altitude experimentation is the investigation of Terrestrial Gamma-ray Flashes (TGFs). Discovered in 1994, TGFs are bursts of gamma rays in Earth's atmosphere thought to be caused by electric fields produced above thunderstorms. Recent observations show that approximately 50 TGFs occur each day – a number that may be much higher due to the possibility of flashes in the form of narrow beams that would be difficult to detect or the possibility that a large number of TGFs may be generated at altitudes too low for the gamma-rays to escape the atmosphere. To investigate this possibility, suborbital access is ideal, since it permits frequent measurements, tailored launch windows, and flexible flight plans, allowing scientists to hand-pick the storms they want to study.

Another group of researchers eyeing the possibilities offered by suborbital flight are planetary scientists such as Alan Stern and Dan Durda, who are interested in dust aggregation and condensation in early planetesimals. It's a process that can be studied in suborbital flight because inter-particle gravitational forces are negligible and collision speeds are low (parabolic flight isn't much good because there is limited time and large accelerations in the "pull-up" phase). Another obvious planetary science mission is in the field of space-based observation. It's why the Planetary Science Institute (PSI) and XCOR Aerospace have signed a memorandum of understanding that lays the groundwork for flying the human-operated Atsa Suborbital Observatory aboard XCOR's Lynx spacecraft (Figure 6.5).

The Atsa project will use crewed reusable suborbital spacecraft equipped with a specially designed telescope to provide low-cost space-based observations. The Atsa

Figure 6.5 Lynx spacecraft carrying the Atsa Suborbital Observatory. Courtesy: XCOR

(which means "eagle" in the Navajo language) observatory will be used to observe Solar System objects near the Sun that are difficult to study from orbital observatories such as Hubble and ground-based telescopes. To ensure optimum data acquisition, the Lynx spacecraft will fly to space on a customized flight trajectory and will be capable of precision pointing, allowing the Atsa system with its operator to acquire the desired target and make the planned observations.

In a related field, heliophysicists are also looking to use suborbital vehicles as a platform to study high-frequency waves and Doppler shift in the solar chromospheres. A similar project flew on the Shuttle but, with the retirement of the Orbiter, human-in-the-loop opportunities no longer exist. Now, with suborbital access on the horizon, heliophysicists are planning on flying special telescopes and solar pointing systems to demonstrate UV imaging from the platform. Once the UV imaging has been proven, the heliophysicists plan to raise the technology readiness level (TRL) of their equipment and fly a new set of detectors.

Panel 6.2. The stand test

Even during a short suborbital flight, the body will undergo some changes that may make the transition back to Earth's gravity a little challenging for some. For example, some crewmembers may find it difficult standing upright due to orthostatic intolerance. Such a simple task may prove problematic because it is in this position that gravity is exerting most of its influence on the cardiovascular system. Think of an upright human body as a tall column of water. As gravity is pulling down on that column of water, each level, or depth, of water is influenced. In any body of water, the pressure at the surface of the water is equal to atmospheric pressure, but the pressure rises 1 mmHg for each 13.6-mm distance below the surface. This pressure results from the weight of the water above it and is called hydrostatic pressure. Hydrostatic pressure also occurs in the vascular system because of the weight of the blood in the vessels.

Hydrostatic pressure exists on Earth because of gravity; when a passenger spends time in suborbital space, their cardiovascular system will try adjusting to functioning without gravity. For those with a poor cardiovascular system, the return to Earth's gravity may prove challenging. To test passengers' orthostatic intolerance, life scientists have suggested administering the "stand test" pre and post flight (the pre-flight measurements will serve as baseline control measurements from which to compare the post-flight data). The measurements may give scientists an indication of what may be happening in the body to cause orthostatic intolerance and help them develop countermeasures. The stand test consists of a 29-minute supine period during which the passengers will be instrumented to measure HR and blood pressure cuff. Next, the passengers will be asked to stand for a 10-minute period, during which measurements will be taken again.

While there are dozens of experiments to keep atmospheric scientists, heliophysicists, and planetary scientists busy for years in suborbital space, perhaps the discipline with the greatest potential is space life sciences, especially in the field of human physiological performance. Space life scientists have already put together a shopping list of the in-situ monitoring tests (Panel 6.2 describes one such test) they plan on conducting, ranging from measuring standard parameters such as heart rate, stroke volume, blood pressure, and oxygen saturation, to less obvious parameters such as measuring brain waves and analyzing neurovestibular and sensorimotor responses. Another research area scientists are particularly interested in is the effect of G-loading from repeated exposures. Unlike government-sponsored spaceflight, which has flown a small population of astronauts with very similar medical histories, scientists studying suborbital passengers and crew will have access to a large population with all sorts of medical backgrounds; some will smoke, others will have high blood cholesterol, and some will be on medication. Of course, such a diverse population has its challenges, but at least it will provide an opportunity for doctors and scientists to understand what can go wrong. Given such a large population with such varying medical histories flying on vehicles with different trajectories and ascent and descent vectors, it would make sense for there to be an industry-standardized central database; hopefully this can be achieved through a cross-industry agreement between providers and the research community.

THE SUBORBITAL PAYLOAD AGENT

While flying so many people into space may prove a boon for space life scientists, there are some who question the wisdom of mixing experiments and tourists, although, for some biomedical applications, tourists may *be* the experiment, as researchers measure their adaptation to weightlessness and back. However, for heliophysics and planetary scientists, tourists could be a pain; an astronomer trying to set up his payload probably won't take too kindly to a tourist crashing into him! The answer? Well, one solution is for a third-party organization to arrange science flights. These "payload integrators" would liaise with the scientists and operators, handle the technical details of integrating the experiments on the vehicles, and ensure the experiments on each flight are complimentary. This might lead to "science charter" suborbital flights, where a vehicle is reserved exclusively for scientific applications. Now, you might wonder why scientists can't use the travel agencies that the suborbital companies are using to interface with customers and coordinate sales activities that way. Well, using existing travel agents probably won't be effective because the cargo customer base is focused on scientific research instead of human entertainment. Of course, there is the option of developing the sales/customer service talent in house, but this would be expensive and distracting from the firms' focus of flying rockets. So, operators launching cargo and scientists will probably need their own class of travel agent, their own out-sourceable sales force, which would be a Suborbital Payload Agent (SPA).

The SPA would be responsible for interfacing with the customers (scientists who want to accompany their payload or those who simply want their payload flown on a flight rack), integrating experiments into flight racks, and delivering integrated flight racks to launch operators for flight. After each flight, the SPA would deliver payloads and data back to the customer. Basically, all the coordinating of payloads, launch logistics, and flight payment would be done by the SPA. An SPA would be a huge benefit to the customer, not only because it would relieve them of all the paperwork we mentioned before, but it would also create more flight opportunities. How? Well, imagine if a customer wants to fly on a particular flight but the manifest is full of tourists. Not a problem, because the SPA will have signed agreements with multiple launch operators so, in the event of a full manifest or, God forbid, a crash, the payload would simply be switched to alternate providers as required. Also, as the commercial suborbital market becomes more competitive, the SPA could leverage their buying power when purchasing flights from the launch provider and the savings could be passed on to the customer.

The SPA would also be beneficial to the launch providers because it would allow them to focus on their core competency – launch operations. Also, outsourcing sales and customer service responsibilities and the associated costs both in time and money would be beneficial, so the SPA would be a godsend and, even if an operator decides to develop a sales force of their own, an SPA would still allow an operator to increase demand at a lower cost.

PAYLOADS

So, what sort of payloads are we talking about? Well, one of the payloads will certainly be human subjects, but we've discussed those. Another popular cargo will probably be CubeSat-sized payloads. A CubeSat (Figure 6.6) is a miniaturized satellite that has a volume of one liter (10-cm cube), weighs no more than 1.33 kg, and usually uses commercial off-the-shelf (COTS) components. California Polytechnic State University and Stanford University developed the CubeSat in 1999 to help universities to perform space science and exploration. Today, the majority of CubeSat development comes from academia, although aerospace giant Boeing has also built CubeSats.

Thanks to their relatively small size, CubeSats can be manufactured and launched[1] for about $65,000–$80,000 – a price tag far lower than most satellite launches,

---

[1] One of the earliest launches of a CubeSat was on June 30th, 2003, from Plesetsk, Russia, with Eurockot Launch Service's Multiple Orbit Mission. On October 27th, 2005, a Kosmos-3M launch vehicle carried three CubeSats into orbit on the European Space Agency's Student Space Exploration & Technology Initiative (SSETI) mission. Two years later, seven CubeSats were launched as secondary payloads on a Dnepr rocket. These included a Colombian project from the students at the Universidad Sergio Arboleda. Their satellite, called Libertad 1, was Colombia's first. Then, on December 8th, 2010, several CubeSats were reported to have deployed successfully from a SpaceX Falcon-9 rocket.

Figure 6.6 An example of a CubeSat. nCube was a series of two Norwegian satellites, built by students at several Norwegian universities and university colleges. Both satellites were built to the CubeSat pico-satellite standard, which allowed one or more cube satellites to be launched by "piggybacking" with a larger satellite. The goal of the nCube satellites was to stimulate interest in science and increase competence in space technology among students and educational institutions. The second goal was to communicate with amateur radio ground stations and to test a space-born AIS receiver for tracking ships and reindeer. Courtesy: Wikimedia

which is why it's a favorite option for schools and universities across the world. It's also why a large number of universities and some government organizations around the world are developing CubeSats and why they will undoubtedly be popular on board suborbital flights.

Another likely payload is the NanoRack (Figure 6.7), an experiment platform built by NanoRacks LLC. Two NanoRacks are installed on the US National Laboratory on board the ISS, each of the platforms holding up to 16 NanoRack payloads in the form of a CubeSat. NanoRacks LLC call the payloads NanoLabs; others call them CubeLabs™. Each payload is a compact 10 by 10 by 10 cm. In

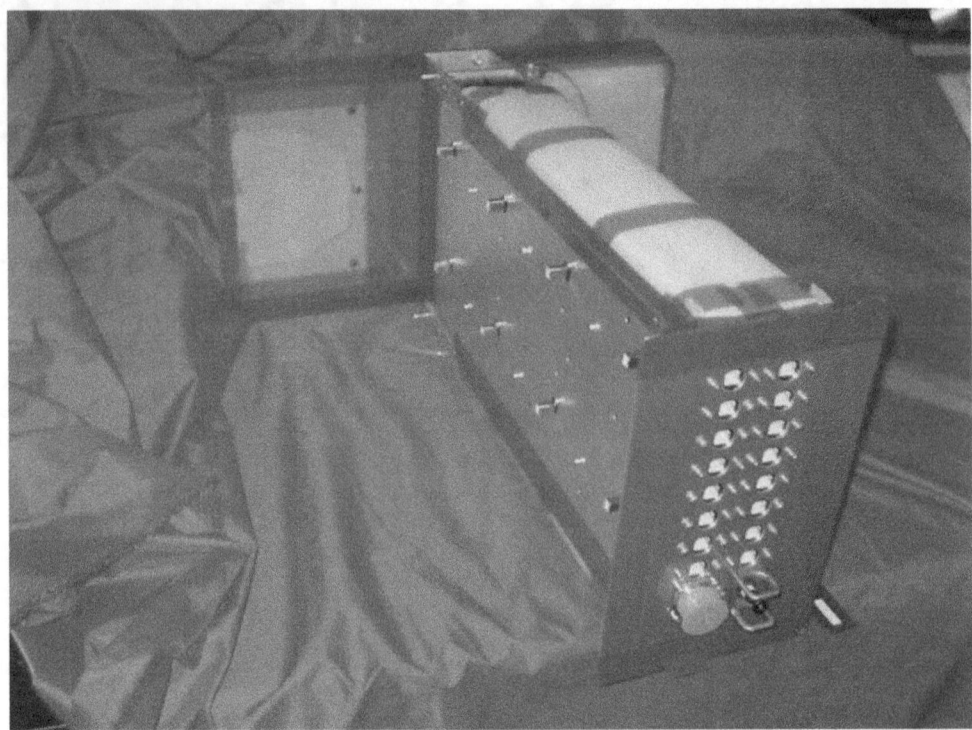

Figure 6.7 A NanoRack. NanoRacks allow small standardized payloads to be plugged into various platforms, providing interface with the Space Station power and data capabilities. The company, led by Jeffrey Manber, brings together entrepreneurs, scientists, and engineers, and has partnerships with several world-class organizations. Courtesy: www.nanoracks.com

addition to being compact, the platforms allow for an easy "plug and play" interface, in which every project on the platform plugs into the space station power and communications system; a similar approach could be used on board suborbital vehicles. The attraction of the NanoRacks platform is that it standardizes what has been, until now, custom-designed hardware: the research platform and payload hardware. By providing a set standard, the barrier to entry for suborbital research will be lowered. Also, advances in miniaturization mean that real science can take place in smaller containers. Thanks to the standardized hardware, the NanoRacks can be manifested and flown within a year for a price that can be as low as $25,000. The price includes the paperwork required for space transportation, the safety review, space transportation, insertion of the payload into the NanoRacks platform, power, and two hours of consulting on payload development.

GETTING SUBORBITAL RESEARCH READY FOR LIFT-OFF

When Alan Stern decided to organize a meeting in 2010 to explore the research applications of the new generation of commercial suborbital vehicles under development, there were many who warned him he would be lucky to get 50 people to attend. Well, those naysayers were wrong. More than 250 people filled the meeting rooms for the first NSRC in Boulder, Colorado. The following year, people once again voted with their feet and attended NSRC by the hundreds. Attendees listened to an amazing variety of proposals to exploit the capabilities of next-generation suborbital spaceflight; in addition to featuring more than 120 presentations spread among 20 technical sessions, the 2011 edition of NSRC featured four discussion panels, a press conference, presentations or booths by 25 sponsors, and a public evening presentation by Virgin Galactic CEO George Whitesides. Clearly, the potential of having access to space similar to that provided by some sounding rockets, but for an order of magnitude lower cost and on a much more frequent basis, has attracted researchers in large enough numbers for suborbital science to be a viable enterprise. Even for those experiments that can't be done on suborbital spacecraft, those vehicles can still provide an opportunity to test equipment intended for later use on orbit. Also, suborbital flights can help engineers and scientists increase TRLs of flight-qualifying hardware that may be difficult or impractical to do otherwise, since suborbital flight will provide orbital quality assurance. It's no surprise, then, that suborbital companies are already examining the suborbital science market. XCOR Aerospace is developing its Lynx vehicle to support experimental payloads located either in the interior of the spaceplane or on various locations on its exterior, including a cargo pod originally designed to host an upper stage for launching nanosats.

Who will be involved in getting suborbital science off the ground? Well, NASA will continue to play a role through its CRuSR program, which will help support research programs that could fly on commercial suborbital vehicles. By supporting suborbital research, the agency will help to promote the development of the industry that could provide NASA with benefits in the future, particularly orbital benefits; after all, many future orbital capabilities will likely evolve from today's suborbital ones. By funding suborbital development, NASA puts itself on a sustainable, step-by-step path to build an industry that evolves to low-cost access to orbit and, without a manned spaceflight capability, it's an opportunity the agency can't afford to pass up. While CRuSR won't – at least in the near term – fund researchers and astronauts for hire to fly on suborbital vehicles, it doesn't stop other organizations like SwRI from flying them.

As the industry will prove, suborbital science missions will provide low-cost access to the space environment. Vehicles will support unmanned and human-tended experiments and the industry will leverage private investment by taking advantage of the millions of dollars of private financing of new suborbital vehicles. Also, the fly-on-demand, rapid turnaround, and human-in-the-loop capabilities will enable new types of research and provide new avenues for scientist involvement. The Sun may have set on the Shuttle era, which used to be the symbol of the "can-do" spirit, but that hardly matters to those involved in the commercial spaceflight industry – an enterprise that will, just as the Shuttle did, push the frontiers of technological achievement and embody the "can-do" spirit of NASA.

7

Orbital Missions

Almost a century ago, the US Postal Service (USPS) set up its first airmail operation. The USPS had its own pilots, planes, and regular routes, transporting letters and parcels from coast to coast. Once the public was accustomed to national delivery, the US government promptly handed over airmail to the private sector. Commercial airlines, which were being established at the same time, bid on contracts to carry the mail. The work guaranteed the fledgling firms an income long before the public was hooked on flying. The companies, which, for many years, doubled as flying postmen eventually became giants of passenger aviation. For example, by the mid 1920s, the USPS, hoping to develop its transcontinental airmail network between New York and San Francisco, offered 12 contracts for spur routes to independent bidders. Some of the carriers that won these routes eventually evolved into Pan Am, Delta Air Lines, American Airlines, United Airlines, and Northwest Airlines. Today, we are on the cusp of a similar transformation by allowing private companies to take passengers and astronauts for hire into suborbital space and eventually into orbit and beyond. Many are predicting the privatization of space will turn out to be the best way to open up low-Earth orbit (LEO) and beyond to a broader cross-section of humanity. With the Shuttle retired, the US no longer has the means to get its astronauts to the International Space Station (ISS). As a temporary measure, Russia's Soyuz rockets will serve as an extortionately priced (tickets are more than $60 million each) taxi service until the US has a new spacecraft of its own. But, as discussed in Chapter 5, rather than have NASA build new spacecraft, the US is hoping to free itself from Russian dependency by introducing financial incentives to entice companies to build new vehicles. As we've already seen, both established and maverick aerospace companies are eager to get a piece of the action; through NASA's Commercial Crew Development Program (CCDev), there are already proposals for suborbital and orbital space vehicles that could be operational by the middle of the decade.

Of course, the privatization of space doesn't mean NASA is getting out of the manned spaceflight business; it just means the agency is redirecting its focus to deep space exploration and landings on Mars. And, while NASA busies itself with its interplanetary plans, LEO will gradually transform into a viable marketplace rather than just a NASA destination. Of course, as LEO becomes commercialized and low-cost, frequent access to space is realized, a veritable constellation of job opportunities will be created. The workers in this extraterrestrial market will include our astronauts for hire, who will be outsourced to work as regular astronauts, miners, orbital payload specialists, tour guides, construction workers, and a myriad of other jobs. To get an idea of how all these people will be employed, we first have to get a sense of how the near-future orbital infrastructure will evolve.

ORBITAL INFRASTRUCTURE

We'll base our near-future infrastructure on a multi-depot architecture; this is a flexible system capable of serving a myriad of needs ranging from crewed exploration, lunar surface habitation, launching interplanetary probes, and deployment of heavy geostationary satellites. The foundation of this multi-depot system will be two space logistics depots (Figure 7.1). The first of these we'll locate at 28.5° at an altitude of 482 km; the second we'll locate at 51.6° at an altitude of 497 km. The lower inclination allows maximum payload performance for vehicles launching from the east coast of the US, while the higher inclination will probably correspond to commercial habitats such as Bigelow's BA-330 module. The space logistics depots will consist of co-orbiting space logistics facilities that will receive cargo and passengers from Earth, provide accommodation, and berth spacecraft for

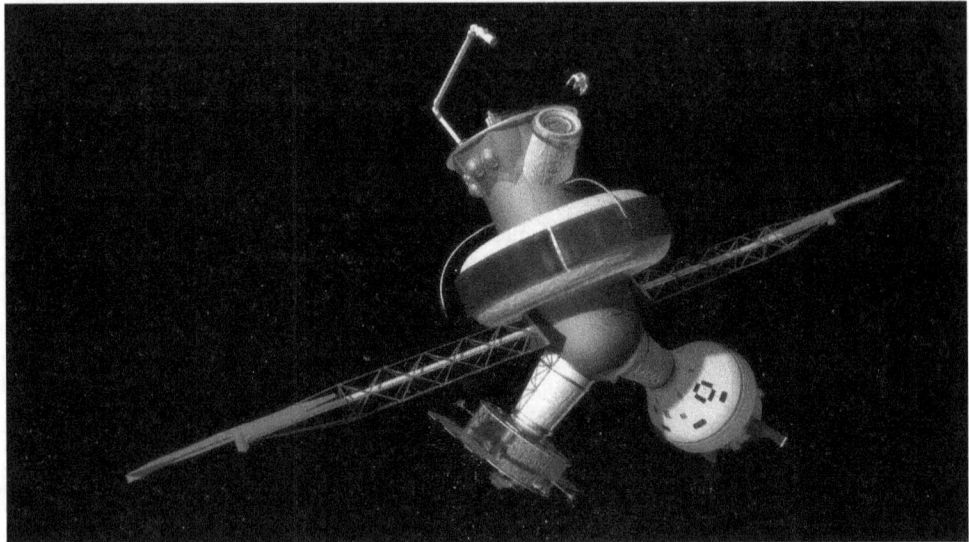

Figure 7.1 Earth–Moon L_1 staging node. Such a gateway, replete with a habitat for crew occupancy, is a good spot to support a locus of activity, such as testing hardware, supporting science operations, and as an astronaut training ground to prep crews for interplanetary missions. Courtesy: Mark Holderman/NASA

Orbital infrastructure 165

Figure 7.2 A science-fiction space tug. Courtesy: www.greywulf.net

use in LEO nearby locations. One of the facilities co-located at the space logistics depot may be a construction station while other facilities may include a research module and a medical facility. To service these space logistics depots and to transport workers will require permanently space-based reusable space logistics vehicles – the space tug and the space ferry.

A space tug is a spacecraft that will be used primarily for maneuvering spacecraft and other objects in the vicinity of a space habitat or another spacecraft whereas a space ferry – which we'll get to shortly – is a spacecraft that will be used to haul other spacecraft, cargo, or people from one orbit to another or from orbit to Lagrange points and to and from the lunar surface. Because space tugs (Figure 7.2) will be tasked mostly with proximity operations, they will probably have one or more robotic arms for better handling, grappling with, and berthing target spacecraft. It won't carry much propellant because it won't have to travel big distances or need to get anywhere in a hurry, although the delta-V capability will probably be in the low hundreds of meters per second range.

Now, the near-future space tug, which will most likely be piloted by our cadre of astronauts for hire, probably won't look much like the vehicle depicted in Figure 7.2. However, it will have a crew module capable of accommodating a crew of two or three for three or four days. It will also be capable of operating autonomously and retrieving sizable payloads from a 185-km circular orbit and returning to the LEO

Figure 7.3 The NeoMiner. Courtesy: Jan Kaliack

space logistics depot. Conceptually, the space tug may look like the vehicle depicted in Figure 7.3. It will be about 10 m long with a diameter of about 3.5 m and an empty mass of around 10 tonnes. As you can see, the vehicle is fitted with all the necessary equipment to conduct its missions: a universal docking adapter to attach itself to other spacecraft, a robotic grappling arm, an airlock for passenger transfers and extravehicular activity (EVA) support, and an equipment bay for transporting materials.

Once the space tug(s) has helped establish the space logistics depots, the next phase of expanding the LEO infrastructure will be to extend space access around LEO and beyond. Since the space tug, versatile as it is, would be unsuitable for this mission, we need to use a different vehicle: the space ferry, or, to use the correct space terminology, a reusable transfer vehicle. The space ferry, in contrast with the space tug, will most likely be a high-performance spacecraft with a delta-V in the order of 4–8 km/sec. The space ferry is a concept that has a history that stretches back to the late 1970s when NASA began investigating space-based reusable systems to provide mobility in LEO. In fact, Boeing conducted a conceptual orbital transfer vehicle (OTV) study that featured in the 1986 "Pioneering the Space Frontier" policy report. Boeing's OTV would transport crews and equipment from space stations in LEO to lunar orbit – a function that led to the OTV being renamed by NASA the Lunar Transfer Vehicle (LTV) to stress its importance for manned lunar base missions.

Since the Boeing OTV used 1990s technology, we'll need to upgrade the space ferry design for use in our notional orbital infrastructure. It may comprise a three-person crew module, a ferry compartment, and a propulsion module. The whole vehicle may be about 20 m in length and would be launched without propellant or payload. In between missions, it would be based in a hangar (which would provide shelter from micrometeorites) at one of the depots, where it would be loaded with consumables and passengers. Flight operations for a typical LEO to Lagrange point/ GEO station mission would be flown at a delta V of about 4 km/sec, while return trips would require an aerobraking maneuver to reduce the velocity to near LEO circularity, a circularization burn into LEO, and docking at the depot.

Later, once the space logistics depots have been established, and the space tugs and space ferries have proven their capabilities, it's likely that a space-gate will be located at Lagrange Point 1 (L1).[1] We know that Bigelow Aerospace has its sights on establishing space habitats at L1 partway between the Moon and Earth. The reason? Well, the L1 location, or first Earth–Moon Lagrange point (EML-1) as it's often referred to, offers a number of advantages that make it an ideal destination for activities in cislunar space. An expandable or inflatable space-gate (Figure 7.4) deployed at EML-1 would be an important stepping-stone for travel to the Moon (another Bigelow destination, incidentally) and beyond and would ultimately serve as the centerpiece of a multi-destination flexible exploration strategy. In addition to serving as a waypoint en route to destinations beyond LEO, these space-gates would serve as a platform to demonstrate next-generation systems necessary for missions beyond the Earth–Moon system. These long-duration-enabling beachheads may eventually provide opportunities for other missions such as the assembly and upgrade of complex science facilities and support for space depot systems.

In addition to the L1 space-gate, another one will probably be located at L2. The L2 (Figure 7.5) space-gate will serve as a propellant depot; it's in deep space but that means it's far away from any planetary surface and so the thermal, micrometeoroid, and atomic oxygen environments are vastly superior to those in LEO. It's also an energy-efficient location, since its position allows the use of powered gravity to assist exit burns (also called Oberth maneuvers), which allow easy access to the lunar surface.

With the space logistics depots, in-orbit refueling capability, space ferry, and L1

---

[1] The Lagrangian – so-called after the Italian-French mathematician who pointed it out, Joseph Louis Lagrange – points (also known as Lagrange points or libration points) are the five positions in an orbital configuration where a small object affected only by gravity can theoretically be stationary relative to two larger objects (such as a satellite with respect to Earth and the Moon). The Lagrange points mark positions where the combined gravitational pull of the two large masses provides precisely the centripetal force required to rotate with them. They are analogous to geostationary orbits in that they allow an object to be in a fixed position relative to a body rather than any more general orbit in which its relative position changes continuously. If the distance is just right – about four times the distance to the Moon or 1/100th the distance to the Sun – a spacecraft will keep its position between the Sun and Earth.

168   Orbital Missions

Figure 7.4 Artist's illustration of a space-gate. Courtesy: John Frassanito and Associates/NASA

Figure 7.5 Rendition of an on-orbit fuel depot concept. Courtesy: NASA

and L2 space-gates established, the infrastructure would be ready to support more commercial ventures but, by this time, technology would be mature enough to consider sourcing construction materials from space rather than hauling everything up from Earth. Even with the development of low-cost vehicles, launching material into LEO will remain a costly exercise, which means if the goal of building an orbital infrastructure is to be realized, another approach must be adopted. One option is to capture near-Earth asteroids (NEAs). It's a job that will require a rather different type of space tug, perhaps like that depicted in Figure 7.3?

So, what will the job of capturing an asteroid involve? Well, first of all, a scout, with a prospector in tow, will be deployed from the space logistics depot. The scout – there will probably be a fleet of them – will travel to potentially interesting NEAs and scan them with sensors to determine chemical, physical, orbital, rotational, and gravitational properties. After completing its sensing activities, the scout will deploy a prospector to land on the asteroid and perform an on-site analysis and dig for samples. Then, job done, the two autonomous spacecraft will return to the space logistics depot, where scientists will analyze the potential of the NEA for mining. Once a NEA has been identified, the scientists will hand the work over to the space tug pilot, another of our orbital astronauts for hire. The space tug will rendezvous with the asteroid and align the landing area with the vehicle's center of mass. Depending on the location of the asteroid, these two phases might take several weeks. Then, the space tug will grapple the asteroid and apply a delta-V by nudging the asteroid (we'll assume scientists have solved the problems of dealing with the asteroid's angular momentum) along a trajectory that will insert it into LEO. Once in LEO, the miners will go to work. First, they will deploy a digger onto the surface of the asteroid that will dig in and break up the material enough for transport to a processor unit. This work will have to be done carefully to avoid stirring up a cloud of debris in orbit around the asteroid due to the low escape velocity. As the digger digs into the asteroid, a transport unit will deliver raw material to the processor, which separates the iron, cobalt, volatiles, nickel, and slag by baking and sintering. As the processor does its job, the extruder system, a sub-unit of the processor, will tailor metal shapes for use in construction in LEO. The metal parts will then be sorted and stacked on the outside of the processor for pick-up by the space tugs. The volatiles will be processed to make propellants for station-keeping and fuel for the space tugs and space ferries, while the unprocessed material may be used for shielding against radiation. Now, this mining business will no doubt be a multi-year enterprise requiring a considerable workforce who will likely live on board LEO inflatable habitats and be ferried to and from their worksite by the space ferries. Gradually, a depot-based infrastructure will be established in LEO, supported by space tugs and space ferries. This infrastructure will eventually be built up in the geostationary and lunar orbits, which will then allow relatively inexpensive access to the lunar surface. So, now that we've established what the near-future in-orbit infrastructure may look like, what about the jobs?

Figure 7.6 Three Sundancer modules linked together in space station configuration. Courtesy: Bigelow Aerospace

EXTRATERRESTRIAL EMPLOYMENT

The new era of extraterrestrial employment began when Bigelow Aerospace became the first commercial company to advertise for professional astronauts. The job advert clearly stated that the company wanted spacefarers with space experience – a requirement that limited the pool of possible of applicants because only 500 or so humans have ever flown in space and many of those have retired! So, what will Bigelow astronauts be expected to do as part of their job description? Well, probable space duties will include working on board the Bigelow Aerospace Station Complex (Figure 7.6) and helping Bigelow's sovereign clients with their payloads and experiments – we'll get to that shortly. This new corps of astronauts will also be expected to manage Bigelow's clients' on-board safety and maintain the cleanliness and operational state of the space station interior; basically, this will mean making sure the life-support system works, fixing things that break, and generally working as a jack of all trades – much like regular government-employed astronauts, come to think of it! Housekeeping and chores? Doesn't sound very exciting, does it? Well, neither was the final mission of the Shuttle era; STS-134's job was to drop off a year's worth of food, equipment, and spares and pick up nearly 5,700 lb of unneeded material – it was essentially a FedEx mission. It should be pointed out that the final Shuttle mission also ran an experiment to see whether a drug cured osteoporosis in mice, which some people might argue was less exciting than delivering the supplies. Leaving aside the argument of whether or not US human spaceflight should have ended with a garbage-collecting mission, no one dreams of going into space to be a FedEx guy; surely there will be more interesting jobs for the aspiring space junkie?

Space traffic controller

Well, yes there will. We've already discussed what the near-future infrastructure may look like and it will be a busy place, with space tugs and space ferries scooting around in LEO and nearby. With all this traffic, there will obviously be a requirement for some sort of traffic regulation. Enter a new breed of traffic controller, who will work at Spaceport Traffic Control (STC). In addition to monitoring, tracking, and de-conflicting inbound and outbound LEO traffic, the space-borne traffic controllers will also be responsible for providing collision avoidance to pilots. To do this, future traffic controllers will direct traffic along space-ways; like today's airways, which enable traffic de-confliction and sequencing as a means of ensuring safe air operations, tomorrow's space-ways will fulfill similar functions. While monitoring these space-ways, the traffic controllers will make sure the space tugs and ferries stick to their assigned routes and re-route vehicles to de-conflict traffic; in the event a vehicle deviates from a route, it will be the responsibility of the traffic controller to report the incident to a controlling agency (possibly the United Nations), which will no doubt issue a penalty against the offender. Now, you may be thinking the job of space traffic controller sounds pretty straightforward, but there are a number of differences between air traffic and space traffic. For example, the speed of most aircraft is about 800 km/hr whereas the speed of a vehicle being launched into orbit will be around 40,000 km/hr! Remember, the delta-V of the space ferry is more than 4 km/sec, so even transiting in local space will require sharp reflexes on the part of the controller if something goes wrong. Then there are the maneuvers and flight paths to consider; compared with orbiting spacecraft, general aviation aircraft are fairly nimble because they can vary their flight paths whereas a spacecraft in orbit follows a fairly deterministic route. Although the space-borne traffic controller won't have to deal with the huge traffic volumes that his/her Earth-bound counterpart does, there are other hazards he/she will have to watch out for: space junk (Figure 7.7), for example. Potential collisions may occur as a result of space debris[2] or uncooperative space systems such as operating satellites that, for some reason, may not be integrated into the space-based traffic control system. From their post at STC, the traffic controllers will utilize space sensors to track flight path degradation and thus avoid collisions. With

[2] There are some 4,000 rocket bodies and satellites, dead or alive, orbiting Earth. In addition, more than 6,000 other large, observable, and tracked bits of debris float around up there. More than 200,000 smaller bits bigger than 1 cm – still potentially dangerous but not tracked – are thought to be in orbit. Much of this material moves at 28,000 km/h.

Figure 7.7 Space junk, also referred to as "space debris" and "orbital debris", is the collection of objects in orbit around Earth that consists of everything from spent rocket stages and defunct satellites to collision fragments. As the orbits of these objects often overlap the trajectories of newer objects, debris is a collision risk to operational spacecraft and orbiting space stations. The majority of the estimated tens of millions of pieces of space debris are small particles, of less than one centimeter. These include dust from solid-rocket motors and surface degradation products such as paint flakes. A much smaller number of the debris objects are over 10 cm. Courtesy: NOAA

improvements in sensor technology, identification and tracking of increasingly smaller pieces of space debris will be possible. Hopefully, by the time a STC is realized, most spacecraft and satellites will be fitted with transponder-like gear, providing constant, cross-linked position updates to the traffic controllers. Eventually, satellites sent into orbit and new launch vehicles will be fitted with internal navigation and housekeeping packages and will be capable of performing (and reporting) routine station-keeping maneuvers on their own, making the job of the traffic controllers a little easier.

In addition to performing standard STC duties, the traffic controllers will probably be tasked with detecting and tracking new launches, and tracking maneuvering targets. Having generated track files and identified vehicles being launched, the traffic controllers will catalog updates in orbit and downlink the information to a central facility providing fusion with other data such as ground-based sensors. Inevitably, there will be occasions when satellites break down in orbit, which will require the deployment of technical troubleshooting staff. The work of these troubleshooters will not only be to resolve satellite emergencies and effect repairs, but also to perform depot-like support for all space systems. Another aspect of the traffic controller's job may be planetary defense; the STC may be outfitted with space sensors to detect potentially dangerous Earth-orbit-crossing objects. Where will the STC be located? Well, a good vantage point will be the L1 space-gate. But, while L1 may be the ideal location for housing the space traffic controllers, most people will probably work in LEO on board the space logistics depots.

Supervisor astronauts

Perhaps one of the first astronaut-for-hire jobs on orbit will be as supervisor astronauts to keep an eye on sovereign[3] astronauts flown by countries lacking an indigenous human space transportation capability; these countries will send astronauts and cargo into space to perform scientific research, acquire technical knowledge, and fly the flag. Between 1978 and 2010, 31 nations without indigenous human spaceflight capabilities have sent 96 astronauts into orbit, and there are plenty more waiting in the wings. Historically, sovereign astronauts have performed missions that can be classified into three basic categories: short-duration visits to space stations, short-duration spacecraft missions such as those performed by the Space Shuttle, and long-duration expeditions to space stations. Based on this historical experience, it's likely the sovereign market will consist of at least two client types; the first type will be interested in short-duration missions (a week to 10 days) to LEO, inflatable habitats, or other commercial space stations for national prestige and scientific and technical research, while the second type may be interested in longer-duration flights ranging from two weeks to six months. The sovereign client may wish their astronauts to be the primary occupants of a habitat or perhaps work as part of a mixed crew with other sovereign astronauts. Either way, supervisor

---

[3] Sovereign customers are already being targeted by Bigelow Aerospace as a key part of its business strategy. Bigelow estimates that 30 flights will be accomplished during the assessment period to support its first operational space station. Since each flight is planned to include three to five passengers in total, that equates to well over 100 astronaut flights over the assessment period. The company has executed seven Memoranda of Understanding (MOUs) with a variety of national space agencies, companies, and governmental entities. These MOUs were signed with organizations in Japan, the United Arab Emirates, the Netherlands, Sweden, the UK, Singapore, and Australia.

astronauts will be required to oversee what will most likely be science and research and technology development activities. Applied research and technology development refers to the use of the microgravity environment and vantage point afforded by in-space platforms to conduct research activities that may lead to commercial application. As discussed in the previous chapter, basic research areas range across the spectrum of biology, chemistry, physics, medicine, and applied research and technology development. Now, sovereign astronauts may be trained scientists but they will only have had basic astronaut training, so it will be up to the supervisor astronauts to make sure everything runs smoothly. To that end, they will be responsible for a spectrum of duties, including inspecting the science work area, ensuring the sovereign astronaut is comfortable, preparing required tools and hardware, and securing them within the work area, inspecting equipment to be used, ensuring power-down of the systems after the science has been conducted, verifying all tools are stowed, and dealing with any off-nominal situation. The supervisor astronaut will also be responsible for reporting problems to Mission Control, and training the sovereign astronauts in the science procedures to ensure crew safety and maintain operability of the on-board systems.

But what about those astronauts not flying under sovereign status – the so-called spaceflight participants or space tourists? If the era of commercial spaceflight has a birthday, it's April 28th, 2001. On that date, American businessman Dennis Tito (Figure 7.8) became history's first space tourist, paying his own way to the ISS aboard a Russian Soyuz spacecraft; in doing so, Tito showed there was money to be made in human spaceflight – potentially lots of money – as he plunked down a reported $20 million for his flight. Since Tito's flight, space tourism emerged as a potentially promising commercial spaceflight market, as evidenced by the people who followed him into orbit. It's difficult to say how many orbital spaceflight participants will fly over the next 10–15 years because the flight rate will depend on the price point. Assuming flights to Bigelow's inflatable habitats are cheaper than the current $35 million ticket price, we can probably expect a dozen or so spaceflight participants to fly. However, if you believe Space Adventures, this market may be much larger; Space Adventures, the company that has brokered every space tourist flight to date, has developed its own forecast of future space tourist flights, which takes into account development of commercial crew vehicles as well as the existence of orbiting space facilities – their forecast calls for approximately 143 passengers flying through 2020. Obviously, these passengers won't be flown to orbiting habitats and simply be left alone; supervisor astronauts will be employed to look after them. They will be responsible for everything from the functioning of the habitat's water supply equipment and food supply subsystem to ensuring there is a steady supply of sanitary hygiene equipment. They will also be responsible for maintaining the life-support system, conducting guidance, navigation, and control activities, docking and undocking, and, of course, troubleshooting. It will be a busy job!

Extraterrestrial employment 175

Figure 7.8 Dennis Tito. Courtesy: Dennis Tito

Orbital entertainment agent

While research and development may be a significant market, voluminous Bigelow-style habitats will no doubt attract the media and entertainment and education markets. If you remember the Mir era, you may also remember the cosmonauts using the Russian space station as a platform to launch promotional activities, most notably ads promoting Pepsi and Pizza Hut; in 1999, Pizza Hut spent $1 million for the rights to plant a large logo on the side of a Proton launcher headed for the ISS. The following year, Pizza Hut worked with Russian food scientists to deliver oven-ready pizzas to ISS inhabitants. Shortly after, Kodak paid to have their logo and a slogan placed onto a material that was to be tested for durability in space on the outside of the ISS. And then, in 2001, Radio Shack & Popular Mechanics worked out deals with the Russians for advertising on the ISS. More recently, Bigelow Aerospace has carried out a "Fly Your Stuff" promotion on board their Genesis I and Genesis II. Photos taken within the Genesis I reveal banners and logos from different companies lined against the module's interior walls.

Spacecraft servicing mechanic

With so many vehicles orbiting in LEO and transiting between gateways, there is bound to be a demand for on-orbit mechanics. This particular cadre of astronauts for hire will be called out to rescue and/or recover stranded vehicles. From reading accounts of on-orbit repair missions, you could be forgiven for thinking that fixing the Hubble Space Telescope is not much different from putting a new headlight in your Toyota Corolla. Well, while the astronauts may make it look effortless, fixing stuff on-orbit is anything but. First of all, there is the zero-G factor, which means that a stray rivet or a broken bolt could ruin a multi-billion-dollar investment. Then you have to consider the fact that the help desk is a few hundred kilometers away. We've already seen several examples of fix-it problems caused by the microgravity environment; for example, during STS-125, when Hubble spacewalker John Grunsfeld opened his tool bag, a stray rivet started rising up in zero-G. Fortunately, Grunsfeld made a grab and rescued the rivet. Grunsfeld's partner, Drew Feustel, didn't fare much better. He had lots of trouble freeing up Hubble's Wide Field and Planetary Camera 2 (WFPC2) for removal. The bolt holding WFPC2 in place wouldn't budge and, when Feustel pushed too hard, his power tool engaged a mechanism designed to avoid putting too much torque on the bolt. He eventually got the go-ahead to override the mechanism, but he had to be careful not to push so hard that he broke the bolt. If that had happened, WFPC2 would have stayed in the telescope, and its $130 million replacement camera would have been sent back to Earth unused. Fortunately, the bolt loosened up with a little extra elbow grease.

Things didn't go quite according to plan for astronaut Heidemarie Stefanyshyn-Piper, either. During her spacewalk outside the ISS in 2008, Stefanyshyn-Piper discovered the grease gun inside her tool bag was leaking, coating everything inside with a film of lubricant. She had begun the job of cleaning and lubricating the gears of the station's malfunctioning starboard Solar Alpha Rotary Joint (SARJ) when she discovered the grease gun leak. Then, while she was trying to clean up the mess, the whole bag floated away. "Oh, great," she mumbled. All Stefanyshyn-Piper could do was look on helplessly while the kit drifted away; it was a bad way to start a six-hour spacewalk! Fortunately, Stefanyshyn-Piper was carrying out the spacewalk with Stephen Bowen, who had his own tool bag with another grease gun, putty knife, and oven-like terry cloth mitts to wipe away metal grit from the clogged joint. Mission Control agreed the spacewalk would continue as planned and that the two astronauts would share tools. Putting her disappointment aside, Stefanyshyn-Piper – the first woman to be assigned as lead spacewalker for a Shuttle flight – completed her work on the joint with Bowen.

Figure 7.9 The interior of an orbital medical facility. Courtesy: Bigelow Aerospace

Space doctors

One of Bigelow Aerospace's many ideas for transforming LEO is the suggestion of joining nine BA-330 modules, three propulsion buses, a docking node, and three crew capsules together to form an Advanced Medical Facility (AMF – Figure 7.9). Such a facility would serve two purposes: to treat patients and to serve as a platform for medical research. But who would the patients be? Well, the construction business is traditionally a dangerous occupation, so these space-faring individuals will no doubt make up a significant proportion of the AMF's patient population. Other patients will most likely be those who have suffered space-related injuries during excursions to the Moon, spacewalks, or perhaps as a result of equipment failure such as a breach of a spacesuit resulting in a rapid or explosive decompression. Caring for these patients will require the first orbital medical staff, including a space-trained physician, an orthopedic physician, a general surgeon, and an anesthesiologist. Two or three advanced Space Medical Technicians (these techs might serve a dual purpose as pilots and lab technicians), a couple of flight engineers, a veterinarian (this facility is likely to serve a dual purpose as a research platform, after all), and ancillary staff will make up the rest of the on-orbit medical team.

Exploration broker

As we'll see in the next chapter, sooner or later, humans will explore beyond LEO. Whether these exploration-class missions will be agency-sponsored affairs or (as is more likely) commercial ventures remains to be seen, but, regardless of which organization embarks upon a deep-space mission, it's likely they will depart from L1, so let's turn our attention there.

L1 will be the place to aggregate mission components for a trip to an asteroid or to more distant destinations. After all, as we'll see in the next chapter, fuel can come from the Moon, while spacecraft elements can come from Earth. Of course, this creates a demand for occupations required to build spacecraft – everything from system engineers to structural analysts and from materials scientists to machinists. And, of course, with all these spacecraft being built, the L1 location will become increasingly busy, requiring a demand for port services, complete with customs officials and orbital tugs, piloted, of course, by orbital transfer pilots. Inevitably, L1 would become a destination in itself – a place where workers and tourists arrive for some R&R between visiting orbital habitats[4] and the Moon. To support the tourist activities, lunar transfer pilots will be needed (preferably ones who can provide comedic banter during flight!) and space travel ticket brokers[5] will help orbital tourists plan their lunar getaways. Presumably, these space travel agents will have flown in space and perhaps visited the Moon so they can give you some insight when arranging your experiential travel. One day, lunar[6] visits will be routine, and you'll be able to simply go online and book the most convenient flight with hardly a thought. But, for the near future, lunar visitors will have to contact their friendly orbital space agent before venturing to the lunar surface.

---

[4] Science-fiction fans who are reading this may remember that Arthur C. Clarke's *A Fall of Moondust* (1961) proposed operations from habitats at the first and second Earth–Moon Lagrange points.

[5] Incidentally, the job of accredited space agent already exists, thanks to Virgin Galactic.

[6] Think this is a bit far-fetched? The world's first space travel agency, Space Adventures, is offering a 17-day trip that will include a stop at the ISS followed by a trip to the Moon's orbit – all for $150 million. According to Space Adventures, the first commercial Moon trip could come as early as 2015. The company has already sold one ticket for the Moon trip and is in talks to sell another one pretty soon. This trip will make the pair of space travelers the first private citizens to see the Moon up close – trendsetters in the grandest sense!

8

One Small Step for Entrepreneurship

Once commercial companies and their astronauts for hire have established an orbital infrastructure, they will probably set their sights on the Moon. It's been nearly 40 years since astronauts last set foot on the lunar surface and momentum is building for a return to establish a permanent manned presence. But the next visit won't be a "flags and footprints" affair. For a start, unlike the Apollo missions, those touching down on future Moon missions won't be government astronauts – they will be employees of corporations, intent on mining the lunar surface for fuels and minerals. You see, the lunar surface is a cosmic cash cow that contains a wide range of precious resources: nickel, silicate minerals, and, of course, fusion fuel – helium-3. So, while it was curiosity that drove NASA's budget for 50 years, the fundamental motivator to return to the Moon will be to create wealth and prosperity, and here's how commercial enterprise will do it.

MINING THE MOON

With all the talk about manned expeditions to Mars and asteroids, the Moon has been overlooked as a destination by the media, but not by commercial space companies. While some analysts may question why we should return to the lunar surface, remember that we're at the beginning of the commercial space era, and commercial companies have to earn money, so returning to the Moon makes complete sense; it's resource-rich with all sorts of minerals and isotopes just waiting to be mined. And besides, with all that wealth hidden in the lunar regolith, mining the Moon could help finance a mission to Mars, and here's why.

In the last few years, the price of helium-3 has increased dramatically from $150 a liter to $5,000 a liter. Most of Earth's stockpile of this isotope is located in an underground bunker near Amarillo, Texas, but that stockpile is rapidly being depleted and could be empty within 20 years. This will be bad for US industry because, without helium-3, there is no prospect of helium-3 fusion, a virtually infinite power source that would help scientists develop a Buck Rogers propulsion system that would get us to Mars in no time at all. Fortunately, the Moon has plenty of

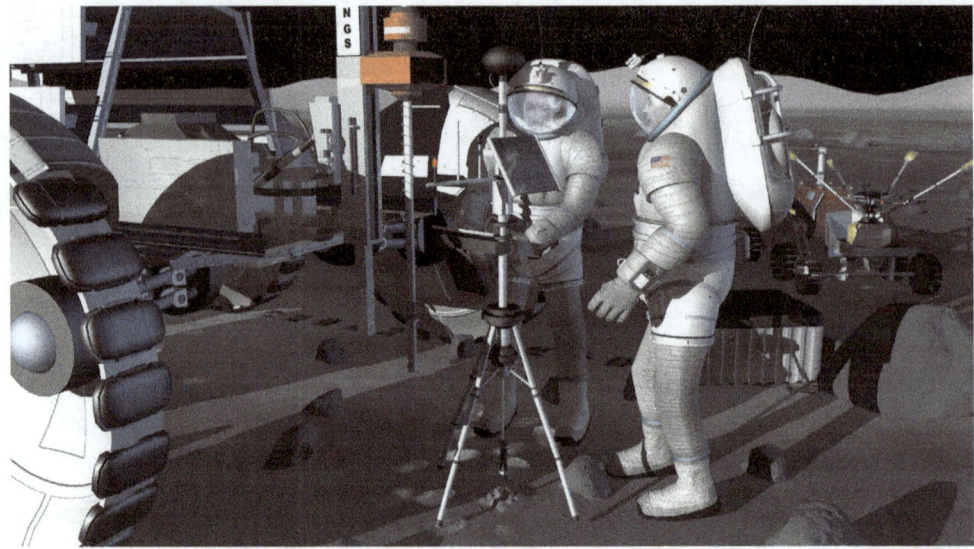

Figure 8.1 Artist's rendering of astronauts engaged in mining activities on the Moon. Courtesy: NASA

helium-3 reserves and, if industry can figure a way to extract the isotope, there's a chance the lunar surface could be the location of a brisk and lucrative import and export trade (Figure 8.1). How will they extract all that helium-3? We'll get to that shortly, but, before astronauts for hire can start mining operations, they will have to establish an infrastructure first, just like they did in low-Earth orbit (LEO).

Infrastructure

First, a series of lander precursor missions will establish resource distribution and characterize the surface before demonstrating technologies and methods for establishing a lunar settlement. Next, a strategic plan will be implemented to use space resources – a step that will include deploying robotic vehicles, landers for dedicated lunar space resource-utilization missions, and lunar surface test beds. Once that's been accomplished, an in-situ, self-sustaining infrastructure of solar energy production and storage will be developed, using lunar materials and wireless power distribution (power beaming). While all these autonomous missions are taking place, test beds on board the space logistics depots, inflatable habitats, and the L1 gateway will be developed for the manned lunar base. These test beds will assess technologies such as closed environmental life-support systems (CELSSs) and in-situ resource utilization (ISRU) systems. The robotic phase will be followed by the first manned mission, which will deploy communications networks (comprising a series of relays and mobile repeaters mounted on autonomous robots deployed to optimal communication locations) and navigation satellite system capabilities for cislunar and trans-lunar space to support lunar

Mining the Moon 181

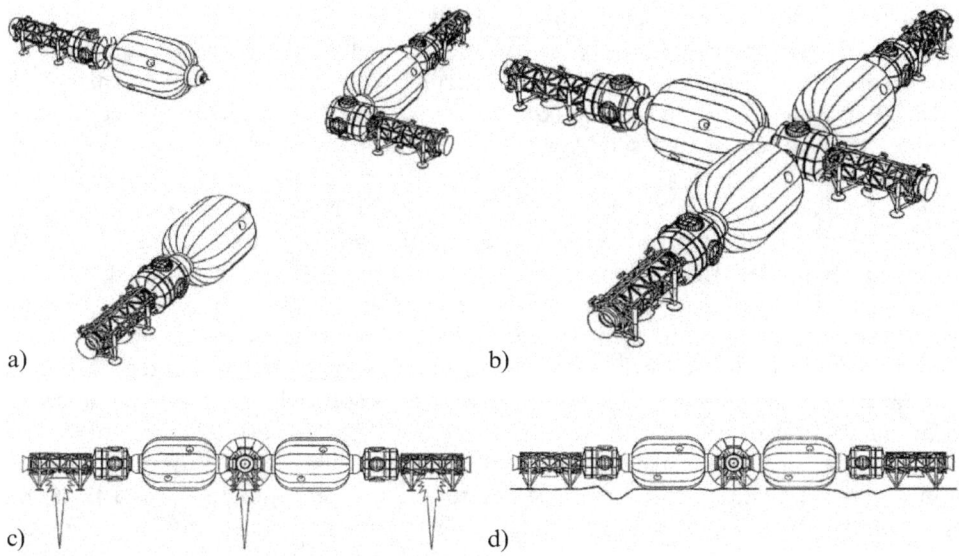

Figure 8.2 Bigelow Aerospace's lunar base. (a) In this phase, the modules are brought together about the central node. Next, a node is connected to each propulsion bus and then to a module, before landing pads are attached to the space busses. (b) In this diagram, the modules are attached to the central node; this constitutes the landing configuration. (c) This diagram shows the assembled habitable structure descending to the surface of the Moon (or some other extraterrestrial mass). The propulsion busses are firing as illustrated by flames. (d) This diagram shows the habitable structure landed on the surface. Courtesy: Bigelow Aerospace

development. The manned mission will also unload a power plant, move it to its operating site, and make it operational. The power plant will be connected to a power-distribution system that will deliver power to an ISRU plant and other surface systems requiring electrical power. Once the power plant is up and running, a thermal control system will be deployed; this will be separate from the power plant, its primary function being to support the other systems needing a means of rejecting waste heat. Once power is on-stream, the crew, who, until this stage have been camping out in the lander, will declare the base "ground operational", at which point they will prepare to receive the lunar base. Since the only commercial company to have designed a lunar base is Bigelow Aerospace, we'll use their mission architecture as a template. As you can see in Figure 8.2, Bigelow's base will be assembled in LEO, where the structure will be robotically constructed from modules, busses, pads, and nodes.

First, two inflatable modules will be placed in LEO and inflated. A central node will be placed in LEO together with propulsion busses and the two modules will be connected to the central node, after which the propulsion busses will be fitted. Landing pads will then be attached to the habitat and the entire structure will be piloted autonomously onto the lunar surface, where it will be hooked up to the power-distribution system; then, lo and behold, you'll have an instant lunar base! If you're thinking building a lunar base is a little far-fetched, there was a similar project dreamt up by none other than legendary Wernher von Braun of Apollo fame. Back in the 1950s, von Braun led a little-known study called *Project Horizon*, conducted by the Army Ballistic Missile Agency. The visionary (and it was visionary – remember, this was more than a decade *before* astronauts had landed on the Moon) project detailed the logistical challenges of constructing and manning a US outpost on the Moon. While it proposed the creation of a *military* base on the Moon, we should be reminded that this was two years before NASA was created and the military was the only place to find rockets. The project planned to construct a small space station in LEO using spent rocket tanks, and the lunar base would have been constructed of pressurized cylindrical metal tanks over a period of three years and 140 Saturn rocket launches! The project went into some detail, selecting potential locations for the outpost based on the most cost-effective orbital trajectories and they even developed a construction schedule. In October 1963, a Saturn 1 rocket would have launched infrastructure material and equipment into LEO and astronauts would have begun constructing a space station with rendezvous, refueling, and launch capabilities. In December 1964, the space station would have been fitted with life support and construction of a second refueling and assembly space station would have begun. The following month, cargo deliveries from the space station(s) to the lunar outpost site would have begun and, three months later, the first two astronauts would have landed at the lunar outpost site. Living in the cabin of their lander, the two astronauts would have verified that the surface environment was satisfactory for an outpost, after which cargo and infrastructure deliveries would have continued. Then, in July 1965, the first nine-astronaut advance construction party would have arrived to begin *Horizon*'s 18-month outpost construction phase. Habitation quarters would have been established and a small nuclear reactor electricity generator would have been placed in protective pits and activated. After six months of construction activities, the *Horizon* outpost would have been almost compete and the crew size would have been increased to 12. Finally, in December 1966, the *Horizon* outpost would have been declared operational, inhabited by a 12-astronaut crew on staggered nine-month rotations.

Sadly, *Project Horizon* never happened, but, today, the project's soul lives on in the heart of Bigelow Aerospace's lunar ambitions. Let's hope they can carry von Braun's torch all the way back to the Moon.

With the habitat on the surface, the astronauts can attend to other outpost tasks (Table 8.1) such as building the landing and launch facilities, excavating regolith for radiation protection, and constructing roads. It won't be easy. For example, without the buffering of air around drilling tools, dynamic friction will be amplified during drilling tasks, generating huge amounts of heat. Drill bits and rock will fuse, hindering progress. If demolition tasks need to be carried out, explosions in the lunar vacuum would create high-velocity missiles tearing through anything in their path, with no atmosphere to slow them down. Then there's the problem of dust, which will settle statically on machinery and contaminate everything. But disturbing dust on the lunar surface is far from being a minor problem; based on NASA's experience with

Table 8.1. Sample crew activity schedule.

Time	Crew	Activity
06:00–06:10	FE	Morning inspection of habitat systems
06:00–06:10	CDR	Comm check with Mission Control
06:10–06:40	All	Post-sleep activities
06:30–06:40	CDR	Morning inspection of habitat
06:40–07:30	All	Morning meal
07:30–07:40	All	Work prep
07:40–07:55	All	Morning planning conference
07:55–08:15	CDR,FE-1	Work prep
08:15–09:00	Two crew	EVA prep: charge fuel cells, plan construction activities and load rover with supplies
08:15–09:00	Team #1 (two crew)	Prepare for IVA activities
09:00–12:45	Team #1	IVA: system monitoring, inspection of systems, sample testing
09:00–09:30	Team #2 (two crew)	EVA crew drive to construction site
09:30–12:15	Team #2	EVA crew unloads, sets up, and operates equipment, and conduct construction activities
12:15–12:45	Team #2	EVA crew drives back to habitat and conduct decontamination activities
12:00–12:15	All	Midday meal prep
12:15–12:45	All	Midday meal
12:45–13:00	All	Afternoon planning conference
13:00–13:45	Team #2	Prepare for IVA activities
13:45–16:45	Team #2	IVA: system monitoring, inspection of systems, sample testing
13:00–13:45	Team #1	EVA prep: charge fuel cells, plan excursion, and load rover
13:45–14:15	Team #1	EVA crew drive to site
14:15–17:00	Team #1	EVA crew unloads, set up, and operate equipment, and conduct construction activities, disassemble equipment if necessary, load rover, and prepare for return to habitat
17:00–17:30	Team #1	EVA crew driver back to habitat and conduct decontamination activities
17:30–18:00	All	Post-IVA/EVA activities (shower, change of clothes, etc.)
18:00–18:30	All	Daily food prep
18:30–19:00	All	Evening meal
19:00–19:30	FE	Evening inspection of habitat systems
19:00–19:10	CDR	Comm check with Mission Control
19:00–19:30	All, except FE	Post-meal housekeeping
19:30–20:00	All	Work prep
20:00–20:30	All	Evening planning conference
20:30–21:30	All	Pre-sleep
21:30–06:00	All	Sleep

the Apollo missions, the biggest contributor to dust generation was the take-off and landing of lunar modules. And this isn't regular dust – 50% of the regolith is smaller than fine sand, which can cause a host of mechanical and health problems: vision impairment, incorrect instrument reading, loss of traction, clogging of mechanisms, abrasion, thermal control problems, and seal failures. It's the reason why astronauts will spend an inordinate amount of time just decontaminating and cleaning equipment.

One of the technologies the astronauts will utilize to the fullest during their construction activities is ISRU. ISRU is a process involving any hardware or operation exploiting and utilizing either natural or discarded in-situ resources to create products and services for robotic and human exploration. For example, on the Moon, *natural* in-situ resources include regolith, minerals, volatiles, metals, water, sunlight, and thermal gradients, while *discarded* in-situ materials include descent stages of lunar landers, fuel tanks, and crew trash. To make the most of ISRU, the astronauts will conduct site surveying and resource mapping tasks to determine usable resources. One of the first ISRU missions the crew will be tasked with will be ensuring their own survival. This will require protecting themselves from the harsh radiation environment by using regolith to bury the habitat. Once the habitat becomes established and ISRU processes become more developed, the crew will begin to process lunar materials to produce water and oxygen, used as a contingency supply for the Environmental Control Life Support System (ECLSS) and as a consumable source for EVA missions. Once they've taken care of protecting themselves and the habitat, they will turn their attention to deploying the mining infrastructure such as regolith excavators, haulers, and surface mobility platforms in preparation for mining helium-3.

MINING HELIUM-3

In 1985, student engineers at the University of Wisconsin discovered that a sample of lunar soil, taken by Apollo 17 astronaut-scientist, Harrison Schmitt, from the rim of the Moon's Camelot crater, contained significant quantities of helium-3. The discovery intrigued the scientific community due to helium-3's unique atomic structure, making it a perfect candidate fuel for nuclear fusion – a process capable of generating considerable amounts of electrical power (the equivalent of a single Shuttle payload – about 25 tonnes – could supply the entire US energy requirements for a year). Non-polluting, and with virtually no radioactive by-product, helium-3 is the ideal candidate as a 21st-century fuel source and there's plenty of it on the Moon; scientists estimate there may be a million tonnes of helium-3, which would be enough to supply the world's energy requirements for several thousand years. Helium-3 is deposited in the powdery lunar soil when solar wind, composed of a stream of charged particles emitted by the Sun, strikes the Moon. It's a process that's continued for billions of years, resulting in a potential cash crop of helium-3 ready to be strip-mined by astronaut miners for hire (Figure 8.3). How will they do it? Well, to extract commercially viable quantities of helium-3, the astronaut-miners

Figure 8.3 Sam Rockwell in a still from the movie *Moon* in which he plays an astronaut-miner. Courtesy: IMDB

will need to process significant quantities of lunar rock and soil. For example, to obtain 100 kg of helium-3, they will have to dig a two-kilometer-square patch of the lunar surface to a depth of three meters. It sounds like a lot of work to extract such a small amount of fuel but, don't forget, this amount of fuel would be worth $140 million and would be sufficient to provide power to a city the size of Dallas for an entire year! Given the value of the isotope, it should come as no surprise that a team of scientists has already acquired the mineral rights for 95% of the Earth-facing side of the Moon, the Polar Regions, and 50% of the far side of the Moon. The team, comprising former NASA astronaut, Dr Joseph Resnick, and NASA consultants, Dr Timothy R. O'Neill and Guy Cramer (the "ROC" team), has secured more than 75% of the Lunar Mineral Rights to allow for the extraction of helium-3. Having secured the mineral rights, the ROC team intend to ensure the Moon does not become scarred by strip-mining visible to orbiting space stations by implementing environmental and visual preservation measures to preserve the lunar surface in the same state as it was prior to mining. To extract their valuable helium-3 harvest, the ROC team will employ one of two strategies. The first, known as the *rectilinear strategy*, is similar to the method used to mine deposits of relatively disaggregated materials on Earth. On the Moon, this process would require that mined regolith be transported by bucket wheel into a mineral processor, in which larger fragments would be rejected by sequential sieving and the finer material would be taken to a volatile extraction unit. Once extracted, the volatiles would be placed into tanks and transported to a refining plant.

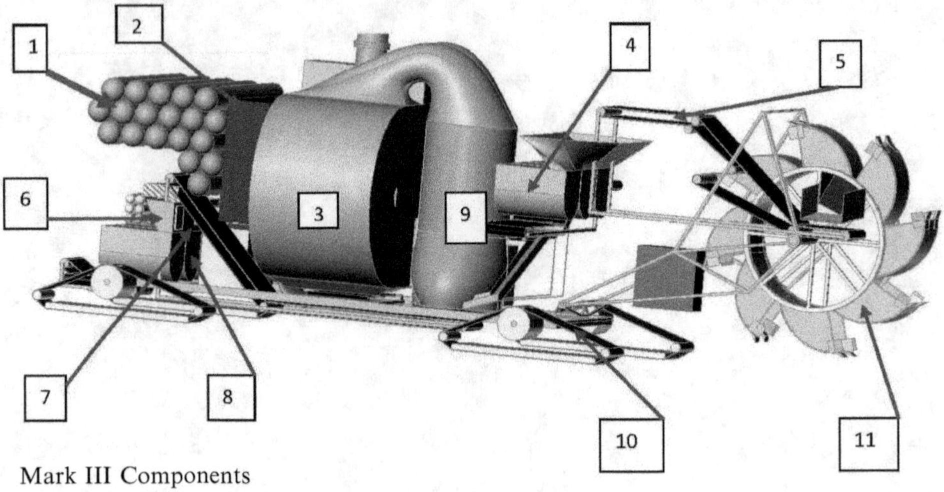

Mark III Components

1. Gas storage tanks
2. Electrostatic separator
3. Heater
4. Screw conveyors
5. Belt conveyors
6. Intercoolers
7. Ejection mechanism
8. Compressors
9. Fluidized chamber
10. Tracks
11. Bucket wheel excavator

Figure 8.4 Mark III Miner. Courtesy: University of Wisconsin, Matthew E. Gajda, Gerald L. Kulcinski, John F. Santarius, Gregory I. Sviatoslavsky, and Igor N. Sviatoslavsky

The alternative strategy, known as *spiral mining*, is a strange fusion of large-scale terrestrial open-pit mining and circular irrigation. Spiral mining requires the placement of a mineral-processor unit attached to the end of a telescoping arm, which, in turn, is attached to a mobile central station. Using this process, the mining and processing would occur in an outward spiral, with volatile materials being extracted from the regolith before being transported to the central station for refining. In this system, the central station would house all the refining, power, control, and habitation elements, which would move together once the telescoping arm had reached its limit of operation, at which point the whole system would move to the next mining area.

Regardless of which process is used to mine regolith, the means of processing it will probably be very similar. Astronaut-miners will most likely use a system called the Mark III Miner, a lunar regolith processor designed by Dr Sviatoslavsky of the University of Wisconsin's Fusion Technology Institute. The Mark III Miner (Figure 8.4), with its arm-mounted bucket-wheel excavator, is designed to "eat" its way through lunar regolith in a 3-m-deep and 11-m-wide trench; mined material flows out of the bucket wheel trays onto a conveyor belt and onto a series of progressively smaller sieves that filter out fine particles. The particles are then transported to the Mark III Miner's pressurized interior. Once inside, the sieved regolith is filtered again by a stream of gas separating fine particles into even finer particles, which are then fluidized, before being transported through a heat exchanger. After being exposed to heat for 20 seconds, the very fine regolith particles are pumped into high-pressure cylinders and transported to the refinery. Once refined, the helium-3, which will now be in the form of either pressurized gas or cryogenic liquid, will be shipped to Earth by unmanned spacecraft fuelled by lunar-derived propellant systems.

MINING OXYGEN

Of course, mining helium-3 won't be the only commercial operation utilizing ISRU. To ensure the evolution of an autonomous lunar outpost, our commercial astronauts will need to produce lunar liquid oxygen (LLOX) utilizing indigenous resources. Not only will such in-situ utilization result in cost savings on propellant for transportation systems; it will also support all those miners living on the lunar surface. Fortunately, oxygen is the most abundant element on the Moon, but the problem is that it must be extracted from lunar rock and regolith, where it exists in chemical combination with elements such as iron and titanium. Energy requirements will also need to be considered because different types of lunar ores require different processes and energy costs to extract the oxygen. For example, one of the simplest processes to produce oxygen is to react lunar minerals, such as ilmenite, with gases but this requires reducing ilmenite with hydrogen by raising the reaction temperature to between 700 and 1,000°C (using hydrogen as a reductant); at this temperature, water would be produced, electrolyzed into hydrogen and oxygen, and the hydrogen would be recycled.

A more cutting-edge oxygen-production process our astronaut miners may use is *carbothermal reduction*. This process combines some of the chemical processes of steel production with the electrolysis and thermolysis of water in which molten reactant is reduced using carbon in various forms. Carbothermal reduction might be used to process simple oxide minerals such as ilmenite, although the very high temperatures required for the process may prove problematic. Another approach may utilize devolatilized carbon to reduce ilmenite, which would require smelting the ilmenite, decarburizing iron, and reforming hydrocarbon. This process would ultimately produce water, which would then be electrolyzed to yield oxygen and recyclable hydrogen.

Another source of oxygen that may be mined is lunar glass. It's one of the most abundant constituents of the lunar regolith, especially in the mare regions, most of which originated as a result of melting produced by volcanic activity and meteorite impacts. These processes resulted in impact glass fusing together rock and mineral fragments into agglutinates, which, in the mare regions, constitute more than half the lunar soil. It's been shown that, by using hydrogen as a reductant, it's possible to reduce lunar glass to iron and water, which can then be easily hydrolyzed or electrolyzed to yield oxygen and recyclable hydrogen. Once again, oxygen might be produced by utilizing the reduction process mentioned previously, although some studies have suggested some agglutinitic glass may be reduced using magnetic separation, due to its high titanium content.

Yet another technique for extracting oxygen is pyrolysis, which requires the application of very high temperatures to induce chemical change and partial decomposition of metal oxides. The temperatures, usually in the range of 2,000 to 10,000°C, are generated by systems such as plasma torches, electron beams, and solar furnaces. Lower temperatures are used for vapor pyrolysis, a process that literally vaporizes the material, transforming it into oxygen-bearing compounds, after which the gas is rapidly cooled and everything except the oxygen is condensed into either a liquid or a solid. One obvious advantage of pyrolysis is it requires no consumables from Earth, since the process is entirely dependent on space resources, such as solar energy and the hard vacuum of the lunar atmosphere.

MINING ICE

With mining systems extracting helium-3 and oxygen in full swing, the astronauts may turn their attention to the ice-rich lunar regolith found in the cold trap areas near the lunar South Pole. Latest analysis of data from Lunar Prospector's neutron spectrometer indicates there may be as much as 300 million tonnes of water ice hidden in the permanently shadowed cold traps near the poles: yet another opportunity for resource-hungry commercial ventures. Dr William Stone certainly thinks so.

Dr Stone, president of Stone Aerospace, is a firm believer that industry, not government, must lead the exploration and mining of the Moon for fuels and water, which is why he's planning to mine lunar water, split it into its hydrogen and oxygen components, and transform it into rocket fuel, before selling it to spacecraft from orbiting gas stations. To achieve his goal, Dr Stone has formed the Shackleton Energy Company (SEC), a wholly owned subsidiary of Stone Aerospace. "Shackleton" refers to the name of the 20-km-wide crater where Dr Stone intends to locate his mining base.

Sure, this may sound like the movie *Avatar*, in which the RDA Corporation mined the mineral unobtanium on the planet of Pandora, but this is no pie-in-the-sky idea. Permanently shadowed craters at both poles have been trapping and accumulating water ice for billions of years, and these stores are a precious resource that could revolutionize space travel; buying gas from orbital gas stations is a lot cheaper than lugging it all the way from Earth. SEC wants to make this happen by first launching robotic scouting missions to the lunar poles; it hopes to be selling propellant in orbit by the end of the decade. In Dr Stone's plan, SEC nuclear radioisotope thermoelectric generator (RTG)-powered rovers will conduct prospecting operations for at least a year, generating ore maps. If the prospecting is successful, the company will continue with human mining missions, deploying a crew of six to eight astronaut-miners for hire. Once the lunar camp is declared operational, human-tended industrial mining and processing operations will commence, leading to frequent transportation of water to LEO processing stations. In these stations, the water will be converted to propellants for point of sale.

SEC's start-up crew will spend a year on the Moon, after which crews will work for six-month intervals. The base will use inflatable structures for the habitat and modular high-power nuclear power supplies for power operations such as heating and electricity. Most of the mining, transportation, and processing functions will be performed by human-tended robots to reduce the requirement for dangerous radiation-soaked extra-vehicular activity (EVA). As fuel demand increases, Dr Stone envisages the base camp expanding incrementally to accommodate increased operations. His customers? Commercial companies offering space tourism, satellite refueling, space repair, and even government agencies. In fact, SEC has already started coordinating with the White House's Office of Science and Technology to drum up support for US nuclear power systems (NPS) to fuel their robotic prospecting systems. Obtaining NPS support will be crucial to SEC's business-planning phase because, without nuclear power, lunar resources like water will not be extractable for decades. Assuming approval is given, SEC can look forward to harvesting a global surface area of over 35 million square kilometers with proven reserves of over a billion tonnes of water ice – enough to launch one Shuttle equivalent per day for over 2,300 years!

SCIENCE ON THE MOON

With a lunar infrastructure established and mining operations providing a steady income, it will be time to start thinking about science. Moon studies will be established to investigate the bombardment history of the inner Solar System and determine the structure and composition of the lunar interior. Scientists will explore the lunar pole, investigate impact processes such as the lateral and vertical mixing of ejecta material, and characterize the lunar surface while the environment remains in a pristine state, which probably won't be that long with all that mining going on! There will also be plenty of opportunities for other scientists; heliophysicists, geologists, and astronomers will all want to take advantage of what the Moon has to offer.

Astrophysics

With astronomy in mind, 194,068 square kilometers on the far side of the Moon have already been set aside for the purpose of establishing a future Moon base for observatories. One of these, the Icarus Lunar Observatory Base (ILOB), will be located at the point farthest away from Earth and provide astronomers with the perfect location to establish optical and radio telescopes, thanks to the absence of atmospheric interference and sunlight. Of course, establishing the ILOB will require lunar construction workers to assemble and emplace the observatory and technicians will be needed to service and clean the optics and remove the ever-present lunar dust, which, charged by the solar wind, will attach itself electrostatically to the components of the observatory.

Astrophysicists will also find the lunar environment an ideal location to perform interferometry and gravitational wave-interferometry, since both are ideally suited to the harsh lunar conditions. By locating radio interferometry antenna arrays on the far side of the Moon, shielded from Earth's radio noise, astrophysicists will be able to observe low frequencies not observed on Earth and may be able to collect data about exotic phenomena such as pulsars, black holes, and planetary radio emissions. Again, it will be commercial astronauts who will be tasked with the assembly of antenna arrays and maintenance activities to ensure components continue to function.

Of particular importance to those living on the Moon will be studies aimed at defining, characterizing, and predicting the lunar radiation environment. Since the Moon is outside Earth's magnetosphere and lacks an atmosphere, it's subject to constant bombardment by energetic solar particles and cosmic rays. By installing large high-energy cosmic ray detector arrays on the lunar surface, astrophysicists will be able to measure radiation striking the surface and perhaps provide a method of forecasting serious radiation events. Once again, the emplacement of such equipment will be the task of our team of astronauts.

The relatively low seismic noise of the Moon may also make it an appropriate location to search for a special, and hitherto unobserved, type of nuclear matter known as Strange Quark Matter (SQM). Although SQM exists only on the pages of textbooks, it's proposed such matter is the most stable form of matter in the universe and may have been produced by the Big Bang itself. Astrophysicists keen on detecting this unique type of matter may install a network of seismometers, since the neutron stars thought to produce SQM often leave behind seismic signatures. If such a signature were to be observed, it would constitute a discovery of profound importance to the field of astrophysics.

Heliophysics

Another means of defining the lunar radiation environment is the study of low-frequency radio astronomy observations, which can inform astronomers of potential solar-particle events (SPEs), which have the potential to kill lunar-bound astronauts and destroy orbiting assets. Due to the ionospheric effects of Earth's atmosphere, it hasn't been possible to perform these observations at frequencies below 10 Mhz but, on the surface of the Moon, there will be no such restrictions; by performing these observations over a number of years, it may be possible to develop a space weather and radiation forecast that will greatly increase the safety of those working on the Moon.

Lunar heliophysics studies may also be instigated to analyze the Sun's role in climate change. To do this, heliophysicists will collect photometric and spectral observations of earthshine in the electromagnetic and charged particle spectrum, which will provide an index of Earth's reflectance. After a month of these observations, it will be possible to calculate Earth's total reflectance, also known as *Bond albedo*, used to determine the amount of sunlight reaching Earth. By comparing measurements of Bond albedo and solar output over many months and years, heliophysicists will gradually develop an understanding of the dynamics of the Sun and Earth's response, which in turn will help them to predict future climate change.

Earth observation

The surface of the Moon is the ideal location for a remote sensing platform, since it affords global observation of Earth. Synthetic Aperture Radar (SAR) will utilize this feature to create highly accurate terrestrial topography, altimetry, and vegetation charts. The lunar surface also provides the perfect vantage point for simultaneous observations of the Earth–Sun system, helping scientists to understand the reaction of Earth's atmosphere to solar activity, in turn helping to estimate the effect of long-term solar changes on climate. The advantage of viewing the whole Earth disc also enables the collation of information concerning land surface mineralogy, land use, land change, and biomass utilization. Observations of the whole Earth disc may also prove useful for monitoring ocean color, sea-surface temperature, and atmospheric studies related to large dust-emission events such as sandstorms in the Sahara. Lunar-based Interferometric SAR (InSAR) will help scientists understand the dynamic response of major ice masses to climate change, as well as permitting the monitoring of sea ice extent and concentration in the Polar Regions.

Geology

One of the first geological investigations may be collating data on Moonquakes, providing lunar inhabitants with a means of determining which areas are safe and which are liable to seismic activity. In addition to revealing site hazards, seismic exploration will also help geologists define mantle characteristics, identify buried lava tubes, and provide an insight into the processes involved in planetary evolution and crustal genesis.

Other geological variables of interest will include characterizing endogenous (from the lunar interior) and exogenous (delivered externally) processes, which resulted in the deposition of volatiles on the lunar surface. Such studies will lead to predictive models of the distribution of volatiles that may assist in the procurement of materials for ISRU. Similar studies may characterize the general geology of the Moon by investigating materials present in regolith to locate and identify elements useful in the development of the outpost.

For geologists interested in impact cratering, there will be plenty of opportunities to investigate and characterize the nature of the heavy bombardment that has occurred during the Moon's lifetime by studying the morphology of old and freshly formed craters and their ejecta distribution. These studies will provide insight into the surface geology of the inner planets and a means of calibrating the ages of surfaces of other planetary bodies, such as Mars. Similar studies will be conducted on meteorite impacts to define the timing and composition of impactors upon the lunar surface. By studying meteorite impacts on the Moon, geologists may gain a better understanding of the impact history of the Solar System, possibly provide clues to the thermal and aqueous history of early Earth, and obtain data supporting or rejecting the existence of past life.

Lunar regolith will also be investigated to help understand the nature and history of solar emissions, galactic cosmic rays (GCRs), and dust from interstellar medium, all of which are preserved in lunar regolith. The value of these investigations lies in uncovering the history of the Sun, since buried regolith may have preserved snapshots of solar radiation events. These snapshots may present geologists with a series of geological time capsules, providing a record of radiation, solar wind, and cosmic ray fluxes – information that may be used to track changes in the Sun. These studies may also be used to support heliophysicists' investigations of the evolution of the galaxy and the history of Earth.

The study of regolith will also permit geologists to characterize the process of space weathering of planetary bodies without air or water. These studies will require mapping regolith maturity to identify regions of ancient and new regolith – data that will not only be used to understand the weathering process, but also provide a reference point for understanding the potential hazards of resource mining on the Moon.

Materials science

One of the primary efforts of material scientists will be to design mitigation strategies aimed to ensure the robust performance of hardware, particularly during extended stays of six months or more. To achieve this, scientists will first study the unique aspects of the lunar environment. These studies will include investigations of fractional gravity, ultra-high vacuum, radiation bombardment, thermal cycling, and perhaps the lunar inhabitants' biggest headache – lunar dust. Characterizing the cumulative effects of the lunar environment will not only permit engineers to make informed decisions regarding the materials best suited to withstand extended use on the Moon, but also provide a useful reference point for mission planners designing Martian mission architectures.

SCIENTIST ASTRONAUT DUTIES

Now, with all these scientific activities going on, the lunar surface will be a busy place, so it will make sense to appoint someone who can coordinate scientific logistics and perhaps serve as a resident technician: a Lunar Support Scientist (LSS), perhaps. The LSS will be an experienced commercial astronaut and scientist who will be responsible for the planned work to be conducted during lunar science missions. Missions will begin with the visiting scientist(s) submitting a lunar time request form to the LSS. This form will contain information needed to determine scheduling requirements and help the LSS prepare for the visiting scientist's mission. The form will also indicate the scientist's first choice of location and time. Once the time request form has been completed, the LSS will schedule transit requirements to and from the lunar surface and determine the cost. Once the cost has been agreed upon, the scientist will submit mission particulars such as citizenship details, visas, vaccinations, medical certification, insurance status, and so on. Once that's been done, a mission identification number will be assigned and the scientist can start dealing with the smaller details such as the logistical coordination of equipment. Before the scientist embarks on their mission, the LSS will send them a prospectus, which will be a summary of the scientific work to be done and a list of the personnel assigned to the mission. On arrival at the base, the LSS will outline work procedures; these will include an outline of the operational schedule, and an orientation session that will include an overview of base policy concerning fire and decompression drills, housekeeping procedures, mess hall protocol, alcohol and drugs policy, smoking, sexual harassment, base safety and accident reports, schedule changes, and communications. Then, the LSS will discuss research plans with the visiting scientists and matters such as access to the base's facilities, availability of technical services and usage of the base's equipment, laboratory, and storage space will be coordinated. A detailed set of notes listing the topics discussed and agreements made will be distributed to the scientists, after which the scientists will be free to conduct their research. During their stay, there may be some scientists who need to ship samples back to Earth or perhaps they will need equipment to be shipped from Earth to the lunar surface. No problem – the LSS will take care of it. Once the scientists have completed their research, the LSS will help the scientists unload their equipment ready for return to LEO or Earth and await the arrival of a new group.

Bio-Suit

At this stage in lunar development, lunar habitants will be spending increasing amounts of time on the lunar surface. To support all this activity, a new spacesuit will be needed. The first spacesuits Apollo astronauts wore on the Moon 40 years ago were designed before humans ever experienced the lunar environment and, while the suits did their job and protected the astronauts, their design left some room for improvement – for a start, they were very bulky and had limited life-support consumables. Remember, one of the key tenets of the return to the Moon will not be to go and check things out and then go home; these commercial missions will last weeks and even months, so the spacesuits will have to withstand a lot more use. Fortunately, research has been under way at the Massachusetts Institute of Technology (MIT) on a Bio-Suit System (Figure 8.5) that incorporates a suit designed to augment the astronaut's skin by providing mechanical counter-pressure. The "epidermis" of this second skin could be applied in spray-on fashion in the form of an organic, biodegradable layer. Incorporated into the second skin would be

Figure 8.5 The revolutionary Bio-Suit. Courtesy: Professor Dava Newman, MIT: Inventor, Science and Engineering; Guillermo Trotti, A.I.A., Trotti & Associates, Inc. (Cambridge, MA): Design; Dainese (Vicenza, Italy): Fabrication; Douglas Sonders: Photography.

electrically actuated artificial muscle fibers to enhance human strength and stamina. The Bio-Suit System could also be integrated with communications equipment, biosensors, computers, and even climbing gear to help astronauts negotiate particularly rugged areas of the lunar surface.

Leading the Bio-Suit research effort is Dava Newman, a professor in the Department of Aeronautics and Astronautics and Engineering Systems at MIT. Assisted by student researchers, Newman's study is targeted at understanding, simulating, and predicting capabilities of suited astronauts in a variety of scenarios, whether performing simple motions or more complex movement, such as overhead or cross-body reach, stepping up, or simply trudging across the lunar landscape. The Bio-Suit-donning scenario envisioned by Newman and her associates is an astronaut first slipping on his or her customized elastic Bio-Suit layer. Then, a hard torso shell would be fitted, sealed via couplings located at the hips. A portable life-support system attached mechanically to the hard torso shell would provide gas counter pressure (gas pressure would flow freely into the wearer's helmet and down tubes on the Bio-Suit layer to the gloves and boots). Thanks to advances in fabrication and application of open cell foam and smart materials like advanced "muscle wire" technologies, the Bio-Suit represents a huge leap forward from the bulky Apollo suits; constructed of spandex and nylon, the multi-layered suit hugs the body's contours like a second layer of skin and its mechanical counter-pressure technology ensures constant pressure is applied to the surface of the body.

Integrated into the Bio-Suit is a bioinstrumentation system that measures biomedical signals such as heart functions, oxygen consumption, and body temperature, while a suite of biochemical sensors provides information concerning body fluids and dosimeters measure the local radiation environment; it's a system that will become increasingly important as surface excursions become longer, EVA locations become more isolated, and activities become increasingly complex, the combination of which has the potential to increase the distraction and fatigue of an astronaut – also, were an accident to occur in an isolated location, having access to the information provided by a wearable sensor system would improve chances of survival.

COMMERCIAL MOON OPERATIONS

Forty years ago, Moon landings were the exclusive province of superpowers but, in the near future, commercial ventures will try turning lunar missions into a profitable business. In the short term, lunar mining operations could supply an outpost with raw materials ranging from solar power, drinkable water, and breathable oxygen to building materials; in the longer run, the Moon could yield beamed energy, fusion fuel, and a wealth of scientific discoveries. Now, some may be wondering why, if there *are* profit possibilities on the Moon, have they been neglected for so long? Well, there are two missing pieces in the commercial Moon puzzle: an absence of a reliable, affordable launch vehicle that can reach the lunar surface and a corps of commercial astronauts. However, with the new breed of commercial vehicles and a new cadre of commercial astronauts waiting in the wings, private enterprise might just succeed where another government-style giant leap might fail. The way we conduct spaceflight is long overdue and, if commercial enterprise chooses to industrialize LEO and the Moon, it will completely transform the aerospace industrial system. In the process, commercial enterprise may relieve Earth of its energy problem, make astronomical amounts of raw materials available, and raise the living standard of people worldwide. The seeds of this transformation are being sown now, as private companies ramp up their spaceflight capabilities and start finding ways to make money in Earth orbit and beyond. In fact, commercial space enterprise is poised on the doorstep of an incredible opportunity to benefit just about everyone. The technology and the ability to make this a reality exist – we need only the will to see it through. Just as we look back and thank those before us for developing things most of us take for granted, such as railroads and highways, the generations to come should be able to look back and thank commercial enterprise for committing to sustainable space exploration. Ten years down the road, people will look back and say: "That was a special time." Too optimistic? Well, let's turn the clock back to 1946 for a moment. Back then, the US Army had formed its Rocket Research Panel, but only a tiny fraction of the nation's astronomers, atmospheric scientists, biologists, and solar physicists appreciated the power that access to space would have on their research. Fast forward 10 years and, lo and behold – rocket-borne research had become so powerful a tool that it formed the centerpiece of space efforts in the International Geophysical Year. Fast forward another 50 years and the commercial space industry finds itself in the same state as the Rocket Research Panel, presented with powerful opportunities that next-generation suborbital and orbital vehicles offer for research, education, and public outreach activities in LEO and beyond. Yet, in a matter of two or three years, one or two suborbital space-lines are expected to be operating commercially. In four of five years, orbital flights will have started and, in 10 years, commercial astronauts may walk on the Moon.

Gradually, as commercial successes mount, as they surely will, the Moon will become a realistically accessible place for other commerce applications. In fact, a sustainable, economically viable commercial enterprise on the Moon won't be limited by the amount of money a company is willing to devote; it will be bounded only by its imagination.

Appendix I

Astronauts for Hire Overview

Founded:	April 12th, 2010
Mission:	To increase the competitiveness of commercial astronaut candidates by providing skills training, facilitating forums for candidate communication, engaging the space research community, and inspiring the next generation
Milestones:	• Joined the Commercial Spaceflight Federation (July 2010) • Hired for first research contract (September 2010) • Assembled Advisory Board (December 2010) • Completed first research flight contract (February 2011) • NSRC 2011 sponsor (February/March 2011) • First A4H training class (July 2011) • Partnerships with multiple research entities (four and counting)
Members:	22 Flight Members from 11 countries speaking 16 languages with 20 MS and 10 doctorate degrees; 10 were recent NASA astronaut applicants (6 advanced to final stages); 11 are pilots, 13 are scuba-divers, and 7 are ZERO-G trained
Capabilities:	Research specialists, payload integration, test, and validation, human test subjects, flight crew (maintenance, operations, mission assurance), product testing, public relations, education, and public outreach

198 Astronauts for Hire Overview

FLIGHT MEMBERS

Christopher Altman

For biography, see Chapter 2.

Ben Corbin, Public Relations Officer

Ben graduated from the University of Central Florida (UCF) in 2008 with a Bachelor of Science in Aerospace Engineering. While there, he worked in the Gas Dynamics Laboratory and the Department of Astronomy as a research assistant and also participated at the International Space University's (ISU's) 2007 Space Studies Program (SSP). He is currently a Ph.D. student in the Massachusetts Institute of Technology (MIT) Aeronautics and Astronautics program, where he recently completed dual Master's degrees in Aerospace Engineering and Planetary Science. He is also serving as the payload specialist for Project VeSpR, a sounding rocket experiment that will launch an ultraviolet telescope into space to study water loss and runaway greenhouse heating on Venus.

Ben is also involved in various space-related extracurricular activities. He designed and built three scientific experiments on microgravity flights and had the opportunity to fly with two of them. He served as the Chief Engineer at the Mars Desert Research Station (MDRS) on Crew 53, where he planned and executed the first ever in-sim construction project. Ben has also written publications on a variety of subjects, including analog spacesuit design, human missions to a near-Earth object and to Mars orbit, flame speed gas dynamics, and planetary hoppers.

Jim Crowell, Chief Financial Officer

James (Jim) is A4H's Chief Financial Officer and a Director of the A4H Board. He is currently an undergraduate student at Arizona State University (ASU), working towards his BS in Earth and Space Exploration. He is the founder and president of the ASU chapter of the Students for the Exploration and Development of Space (SEDS-ASU) and has worked with Dr Phil Christensen, using the Mars Odyssey thermal emission imaging system (THEMIS), photographing and mapping areas of the Martian surface. Jim's also been involved in planning phases, with Dr Matt Fouch and Dr Kip Hodges, of setting up a network of seismometers across the lunar surface to map the subsurface structure of Earth's Moon. Jim's future plans are to continue on after ASU to graduate school to achieve his Ph.D. in a space-related discipline. His interests lie primarily in the colonization of moons and planets.

Jon-Erik Dahlin

Jon-Erik was interviewed in the European Space Agency's (ESA's) astronaut selection campaign and was among the top 2% of applicants. Professionally, Jon-Erik has experience from academic research, education, research administration, and entrepreneurship. His research field as a Ph.D. student at the Royal Institute of Technology in Stockholm, Sweden, was fusion plasma physics and his thesis was based on numerical simulations of turbulence control in a fusion reactor. After receiving his Ph.D. degree, Jon-Erik worked as a program manager at the Swedish Energy Agency, responsible for assessing and initiating research projects and programs, mainly within the area of combined heat and power. He returned to academia, working as a teacher at the Royal Institute of Technology, before launching his current business as a consultant in research, education, and project leading. Besides his professional activities, he is a private pilot and is planning a flight around the world in a small twin-engine airplane. He's also a scuba-diver and has dived in the Indian Ocean, the Pacific Ocean, as well as the cold waters of Scandinavia.

Kristine Ferrone, Fundraising Officer

Kristine Ferrone is a Space Vehicle Concept Design Engineer with The Aerospace Corporation in Washington, DC, where she performs integrative spacecraft systems engineering and mission assurance for government agencies, including NASA and the National Reconnaissance Office (NRO). Prior to working at The Aerospace Corporation, Kristine was a flight controller for the International Space Station (ISS) at NASA's Johnson Space Center (JSC) supporting mission planning and timeline development for crew and ground activities. She was certified in three console positions and was a lead planning engineer for the STS-134 and STS-131 missions to the ISS. Prior to her work in flight planning, Kristine was a Senior Mission Scientist and flight controller for the ISS Payloads Office. She also served as the liaison between real-time ISS operations and the primary investigators for payload experiments on board the ISS.

In 2009, Kristine served as Interdisciplinary Scientist for the 12th crew to the Mars Society's Flashline Mars Arctic Research Station (FMARS), living and working in an analog Mars environment for one month. She completed over 17 hours of Extra Vehicular Activity (EVA), assisting with the deployment of a geophysics experiment, flight of an Unmanned Aerial Vehicle (UAV), and the location of a gypsum deposit used to synthesize water. Kristine also led her own research project at FMARS, studying the benefits of High Power Laser Therapy (HPLT) on treating acute and chronic pain and injuries to crewmembers.

A 2004 graduate of Carnegie Mellon University with a B.S. in Physics/ Astrophysics, Kristine's undergraduate research focused on high-energy particle physics and, from 2004 to 2006, she worked as an Accelerator Operator in the Main Control Room for the Collider-Accelerator Department at Brookhaven National Laboratory (BNL), where she was responsible for the real-time operations and troubleshooting of the particle accelerator complex. While at BNL, Kristine also worked in NASA's Space Radiation Laboratory, studying the effects of space radiation on astronauts and space vehicle equipment.

Not content with just one Master's degree, Kristine has three – a M.S. in Space Architecture with a thesis on command and control concepts for long-duration human spaceflight, a M.S. in Sports Medicine studying exercise physiology and the effects of microgravity on astronauts, and an MBA with a certificate in Entrepreneurship and Technology Management.

Kristine received her FAA Private Pilot License (PPL) in 2011 and has logged over 65 hours of flying time. She's also certified as a PADI Advanced Open Water Diver and, in July 2011, she completed the Suborbital Scientist Training Program (SSTP) at the NASTAR Center.

Amnon Govrin, Chief Technology Officer

Amnon Govrin has a B.Sc. in Electrical Engineering from the Technion University in Haifa, Israel. As an Israeli, Amnon served in the Israeli Defense Force for three years, working as a test and experimentation engineer, inventing and improving terminal ballistic experimentation methodologies. Amnon also wrote experiment decoding software for X-ray imaging and received international recognition for improving precision of behind-armor-debris experiments.

After his military service, Amnon worked for Mercury Interactive, where he developed unique solutions for administration of large enterprise monitoring applications. In 2001, he was sent to Boulder, CO, to spearhead integration of a newly acquired company called Freshwater Software. In this function, he initiated, architected, and developed cutting-edge projects leveraging existing technologies and developing possibly the first AJAX (Asynchronous JavaScript and XML) UI framework used in an enterprise application. In 2004, he started working for Webroot Software, a provider of security software, where, after helping to build the enterprise endpoint protection product, he became a software development manager for the main consumer product line of the company. In that function, Amnon developed and released the 2011 security product line, which received the *PC-Magazine* Editors' Choice Award. Amnon recently moved to Seattle, WA, and now works as Software Development Manager for Amazon Video on Demand.

206   Astronauts for Hire Overview

Melania Guerra, Membership Committee Chair

Melania Guerra is an engineer, oceanographer, and scientific diver. Her doctoral research at the Scripps Institution of Oceanography (SIO) in San Diego, California, focuses on underwater ambient noise and marine mammals. She has seven years of oceanographic experience as a member of the Laguna San Ignacio Ecosystem Science Program, collecting data using acoustic sensors and tags applied with suction-cups on gray whales. She has also performed fieldwork on other whale species in the Arctic Ocean, along the California coast, and in Southeast Alaska. Melania is a certified AAUS scientific diver and has completed cold water/weather survival training and helicopter underwater egress training.

In 2002, before joining SIO, Melania worked at the Advanced Space Propulsion Laboratory at NASA's JSC. While at NASA, she worked under the tutelage of legendary seven-time Shuttle astronaut Franklin Chang Diaz, conducting experiments on a High Temperature Superconductive Magnet, part of the revolutionary VASIMR propulsion system.

Originally from San Jose, Costa Rica, Melania received a B.S. degree in Mechanical Engineering in 2001 from the Universidad de Costa Rica. There, she interned at the Applied Nuclear Physics Laboratory, performing spectral analyses with a residual gas analyzer mass spectrometer. In her spare time, she enjoys communicating science through public outreach programs. She is often an invited scientist at Scripps' Birch Aquarium and at local San Diego and Baja California schools, and has appeared on televised media, promoting oceanographic research.

Mindy Howard

Dr Mindy Howard is an American who obtained her Bachelor's and Master's of Science in Industrial Engineering in the US and a Ph.D. in Industrial Engineering from the Technical University Eindhoven in the Netherlands. She has made the Netherlands her home since 1990 and also has Dutch citizenship.

She began her career at the Royal Dutch Shell Group of companies in The Hague in 1994 and has held several technical and leadership roles in the company. In her latest role, she was responsible for creating strategy and leading implementation globally across four main specialist areas: Sustainable Development, Social Performance, environment, and CO_2 management within the Gas and Power Division.

More recently, she became the Director of the Business and Sustainable

Development Center at the International Space Transport Association (ISTA) in The Hague, where she integrated human factors, sustainable development, and spin-off business ideas in the space environment into the business plan of the start-up organization.

After obtaining her Ph.D., Mindy applied to NASA and was placed on the agency's "Highly Qualified" astronaut candidate list. During her Master's degree, she designed and conducted experiments in fault diagnosis of pilots to facilitate the design of an expert system for the cockpit in a Boeing 737 in conjunction with the NASA Langley Research Center's Flight Management Division (where she won a NASA Graduate Student Researcher Fellowship). During her Bachelor's degree, Mindy worked at Grumman Aerospace, where she assisted in the development of the NASA Space Station Control/Display design for the generic workstation and conducted experiments on tele-operated systems using astronaut subjects. Mindy has also written publications on a variety of subjects, including human factors design of cabins of spaceships, expert systems to help pilots with fault diagnosis, and quality of group decision-making as groups undergo several tasks.

Mindy has completed her SSTP certification at NASTAR, is a certified PADI open water scuba-diver, and has taken an emergency egress course in helicopter ditching, which was necessary to work offshore in the oil industry.

José Miguel Hurtado, Jr

José earned a B.S. with honor and a M.S. in geology from Caltech in 1996 before going to MIT to earn his Ph.D. At MIT, Jose worked on the tectonic evolution of the central Nepal Himalaya and led four six-week expeditions to the Upper Mustang region, covering over 400 km on foot at elevations up to 4,500 m. He also spent a summer conducting structural geologic research in a remote fjord in East Greenland. After MIT, José spent time at the NASA's JPL as a postdoctoral research fellow and worked on various Earth and Mars remote sensing projects. He joined the faculty at the University of Texas at El Paso (UTEP) in the fall of 2002 as an Assistant Professor in geology and was granted tenure and promoted to Associate Professor in 2008. He currently serves as Graduate Advisor and chair of the Graduate Committee in the Department of Geological Sciences at UTEP.

José teaches courses in physical geology, physical geography, field methods, geologic mapping, remote sensing, digital image processing, isotope geology, tectonic geomorphology, and planetary geology. He has expertise in planetary and terrestrial geology, specializing in structural geology, geomorphology, geochronology, and remote sensing, and is funded by NASA grants to characterize the tectonics of the Bhutan Himalayas and volcanic hazards in El Salvador using integrated remote sensing, geophysics, geomorphology, and geologic mapping.

Along with several of his students, José has collaborated with the Astromaterials Research and Exploration Science group at JSC to catalog the geologic activities conducted on the Moon by the Apollo astronauts. An interviewee for NASA's 2009 astronaut candidate program, José is involved in various lunar geologic activities through the Lunar and Planetary Institute, the NASA Lunar Science Institute, and the newly created NASA/UTEP Center for Space Exploration Technology Research. José has served as a member of the NASA K-10 lunar robotic reconnaissance science team, the NASA Desert Research and Technology Studies (RATS) science team, and an instructor for NASA field geologic training courses, including training of the 2009 Astronaut Candidate class.

He is a licensed private pilot and has a long-standing interest in aviation, space exploration, and human spaceflight. In his spare time, José enjoys travel, camping, hiking, running, reading, and aquascaping.

Kris Lehnhardt

Born and raised in Canada, Kris completed his Hon. B.Sc. in biomedical sciences from the University of Guelph in 1999 and obtained his M.D. (2003) and Emergency Medicine specialization (2008) from the University of Western Ontario. He is a Fellow of the Royal College of Physicians and Surgeons of Canada and a Diplomate of the American Board of Emergency Medicine.

Kris started his medical career at the London Health Sciences Centre in London, Ontario, Canada, and is currently working in Washington, DC. He is an Assistant

Professor and Attending Emergency Physician at the George Washington University, where he also serves as the Medical Director for the Emergency Medical Response Group (EMeRG) and as the director of the Wilderness Medicine course.

Kris has trained with the flight surgeons at the NASA Kennedy Space Center (KSC) and at the Defence and Research Development Canada (DRDC). He has completed the SSP at the ISU in Barcelona, Spain (2008), and was a highly competitive applicant and interviewee in the most recent Canadian Space Agency (CSA) astronaut selection campaign. He holds a PPL in both Canada and the US and is a certified open water scuba-diver.

Joseph E. Palaia, IV, Secretary, Business and Legal Affairs Committee Chair

Joe is an entrepreneur, engineer, and technologist whose personal mission is to create the first permanent settlement on Mars and to pursue development in the inner Solar System. He is a cofounder and Vice President of Operations/R&D of the 4Frontiers Corporation, a space technology, entertainment, and education company. He is also co-founder and manager of NewSpace Center, LLC, a firm developing a space-themed research and entertainment facility (INTERSPACE) to be located in Titusville, Florida.

In 2009, he served as executive officer and chief engineer for a one-month simulated Mars mission at the Mars Society's FMARS on Devon Island in the Canadian arctic. While there, he became the first person to launch and operate an Unmanned Aerial Vehicle (UAV) while wearing a simulated spacesuit. He also deployed a prototype lunar rover provided by the Omega Envoy Project, a participant in the Google Lunar X PRIZE, for whom he serves as an advisor.

Joseph holds a B.S. degree in Electrical Engineering from the New Jersey Institute of Technology and a M.S. degree in Nuclear Science & Engineering from MIT. He has played an integral role in two commercial design studies of the first permanent Mars settlement and is co-author of technical papers on the topics of Mars nuclear power plant design, Mars settlement architecture, space economics, and the economics of energy on Mars. He speaks frequently at conferences and events worldwide on the topics of Mars settlement and space development.

Jason Reimuller, Chief Operating Officer

Jason comes from a diverse background, blending scientific research and analysis, engineering, and the operations of aerospace platforms. He is an experienced system engineer and project manager and recently founded Integrated Spaceflight Services, based on experience he gained while conducting various analyses for NASA's Constellation Program. Specifically, these analyses involved abort scenario and launch availability analyses, rescue and recovery trade studies, and contingency crew egress testing. In this capacity, he has worked frequently with NASA astronauts and mission planners to integrate elements of programmatic and technical risk with operability, cost, and technical performance measures to support program-level trade studies.

Jason also gained engineering experience through work with satellite propulsion systems at Space Systems Loral, LiDAR systems at the Sondrestrom Incoherent Scatter Radar Facility in Greenland, and through entrepreneurial endeavors involving aircraft remote sensing payloads and active hydroponic systems. His exposure to spacecraft operations came first as a space command officer in the US Air Force, serving later as a satellite flight director for Space Systems Loral.

Jason completed his Ph.D. in Aerospace Engineering Sciences at the University of Colorado in Boulder. He also holds a M.S. degree in Physics from San Francisco State University, a M.S. degree in Aviation Systems from the University of Tennessee, a M.S. degree in Aerospace Engineering from the University of Colorado, and a B.S. degree in Aerospace Engineering from the Florida Institute of Technology. A recipient of the NASA Graduate Student Research Fellowship, he researches the upper-atmosphere through coordinated observations of noctilucent clouds, where he uses his experience as a multi-engine, instrument-rated pilot to fly research aircraft in Northern Canada to obtain imagery synchronized with satellite overpasses. He also conducts research in glaciology to better understand the dynamic changes of the Greenlandic ice sheet by working with NASA's Operation ICE Bridge, an airborne remote sensing campaign, and integrating the data with space-based and ground-based observations.

Jason is an avid rugby player, a rescue scuba-diver, outdoorsman, aviator, and guitarist. He speaks Spanish and is currently studying Russian. He travels extensively and has been engaged in several international projects. Like several A4H members, he has completed NASTAR's SSTP.

Luís M.R. Saraiva, Chief Science Officer

Luís is a neuroscientist, adventurer, and a space advocate. He is originally from Portugal, where he lived until 2004. There, he completed a "Licenciatura" (equivalent to BSc + MSc) in Biology at the Universidade de Evora and Instituto Gulbenkian de Ciencia, having specialized in Ecology, Paleontology, and Microbial Evolution and Behavior. He then moved to the University of Cologne in Germany, after being selected to be a Fellow of the International Graduate School in Genetics and Functional Genomics. While there, he completed several courses, lab rotations, and his Ph.D. thesis in Genetics/Neurobiology. In 2008, he applied, but was not selected, to be an ESA astronaut. In despair, he moved to Boston, where he worked as a visiting scientist at Harvard Medical School. In October of the same year, he became a Postdoctoral Fellow at the Fred Hutchinson Cancer Research Center in Seattle. He works in Dr Linda Buck's Lab, trying to unravel the neural circuits that regulate innate behaviors, like aggression and fear. He presents his work in conferences and has published several research papers in high-impact-factor scientific journals.

In January 2010, he completed a 14-day expedition to the MDRS in Utah. There, he served as the Biologist and Health and Safety Officer (HSO) of the six-member MDRS Crew 89. He logged nine EVAs conducting his own research projects or helping with other crew members' projects. He also conducted research in the laboratory and worked together with the crew engineer, developing and testing a shower that can recycle water.

Luís enjoys soccer, kickboxing, rock climbing, and mountaineering. He is also a travel and adrenalin junkie and has visited 25 countries. He thinks that our future as a thriving civilization heavily depends on both the "genetic enhancement" of humanity and the space exploration and colonization.

Erik Seedhouse, Chief Training Officer

For biography, see Preface.

Brian Shiro, Chief Executive Officer

Brian's lifelong ambition is to explore space and, in doing so, help improve life on Earth. He is a veteran of numerous geophysical field expeditions to remote locations, including Antarctica, Alaska, Canada, and various tropical Pacific islands. Brian's diverse background includes jobs working in upper-atmosphere physics, carbon nanotubes, satellite radar mapping, geochemistry, glaciology, geodesy, biophysics, impact cratering, seafloor mapping, and high-performance computing. He has experience working at three NASA centers and served as the principal investigator for a Mars Geophysical Lander mission proposal to NASA in 2003. Brian has also completed two analog Mars missions, where he logged 20 EVAs to conduct field science wearing mock spacesuits. These included a 30-day expedition to the FMARS on Devon Island, Canada, in 2009 and a 14-day expedition to the MDRS in Utah in 2010, where he served as Commander of the six-member crew. In 2009, he made it to the "Highly Qualified" level (top 3–12%) of NASA's astronaut selection process.

Brian holds a B.A. with triple majors in Integrated Science, Geology, and Physics from Northwestern University and a M.A. in Earth and Planetary Sciences from Washington University in St Louis. While working on his Ph.D., he left academia to accept a position as Geophysicist with NOAA at the Pacific Tsunami Warning Center in the wake of the 2004 Indian Ocean tsunami. His high-pressure job is regularly featured by news media outlets like CNN and NPR. In summer 2005, Brian attended the ISU Summer Session in Vancouver, Canada, where his concentration was space policy and law. The following year, he gave an invited presentation on his ISU team's work involving global wildfire forecasting to the United Nations Committee on the Peaceful Uses of Outer Space (UNCOPUOS) in Vienna, Austria. In 2010, Brian completed a M.S. in Space Studies from the University of North Dakota with a specialty in human missions to Mars.

Brian is a runner, scuba-diver, aviation enthusiast, environmentalist, Eagle Scout, and outdoor extreme sports enthusiast. He has completed six marathons, including the Boston Marathon. Brian wishes he lived in the mountains, where he could maintain his mountaineering, skiing, and river kayaking skills, but instead settles for living in Hawaii with his wife and two young children.

Laura A. Stiles

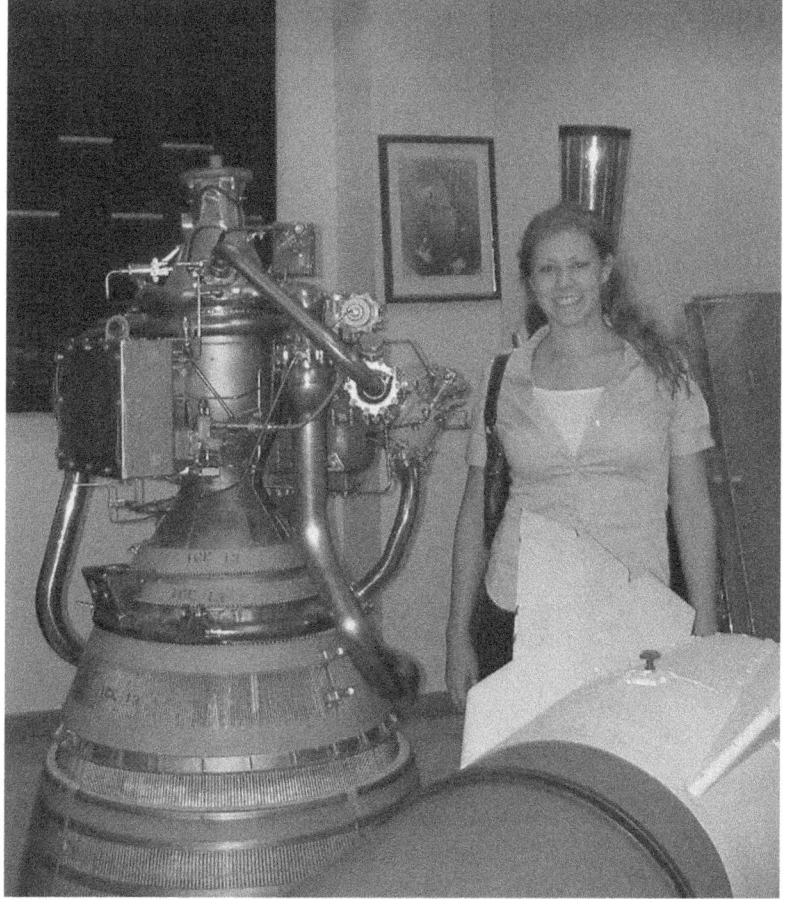

Laura Stiles is currently working toward her Ph.D. in Aerospace Engineering Sciences at the University of Colorado at Boulder. Her focus area is Astrodynamics and Satellite Navigation, and her Ph.D. topic involves studying the use of electrostatic interactions between spacecraft for formation flying or space structures. Laura is a recipient of the National Science Foundation (NSF) Graduate Researcher Fellowship (2008) and the Zonta International Amelia Earhart Fellowship.

Before coming to Boulder, Laura obtained her B.S. in Engineering Physics with a specialization in Aerospace Systems at the University of Kansas. She has served as launch director for high-altitude balloon flights, promoted educational outreach for science and engineering, and was the team leader for a microgravity engineering research project. During NASA's Reduced Gravity Student Flight Opportunities Program in 2007 and 2008, Laura served as an operator on two parabolic flights. Internship/work experiences include positions as a Research Associate with the 2007 NASA Academy at Goddard Space Flight Center, a research engineer at CERN (European Center for Nuclear Research), and a research assistant with the Space Scholar Program at the Air Force Research Lab in 2009.

Laura enjoys spending her free time in the sky as a licensed skydiver or on her snowboard in the powder of the Colorado mountains. She also loves Ultimate Frisbee, swimming, hiking, and anything outdoors.

Alli Taylor

Alli is a New Hampshire native who was inspired by the legacy of Christa McAuliffe from a young age. She is interested in all aspects of space, including human spaceflight, scientific missions, planetology, operations, and policy. Since joining Lockheed Martin and moving to the Washington, DC, metropolitan area in 2008, Alli has learned about constellation planning and collision avoidance analysis of satellite systems. Currently, she is working on her Master of Science degree in Space Studies through the distance program at the University of North Dakota.

Alli gained valuable experience in the commercial space industry while employed at Bigelow Aerospace in Las Vegas as a systems engineer and satellite controller; she spent over two years developing a variety of payloads for the Genesis I prototype inflatable module as well as leading Mission Control operations for both Genesis I and II vehicles. During her time in Las Vegas, Alli became a certified coach for the Zero Gravity Corporation and had the opportunity to fly on two zero-gravity flights.

Alli graduated with a Bachelor of Science degree in spring 2005 from the Florida Institute of Technology. During her senior year, she participated in the Lunar In-situ Resource Utilization (ISRU) team competition, in which the Florida Tech design team was selected as a finalist. During and after undergraduate school, Alli had two internships that were major influences on her career direction. At Ball Aerospace, Alli learned firsthand about satellite systems integration and environmental testing and had the opportunity to work on the Ralph instrument, currently flying aboard NASA's New Horizons mission to Pluto. Alli also interned at the Florida Space Research Institute (FSRI), where she worked with Dr Samuel Durrance on the Station High-performance Ocean Research Experiment (SHORE), a proposed ISS camera payload.

In the near future Alli plans to apply for a Mars analog mission to gain EVA experience. Her hobbies include supporting the Boston Red Sox, drawing, volleyball, roller derby, and snowboarding.

Abhishek Tripathi

Abhishek earned a B.S.E. in Aerospace Engineering in 1998, a M.S. in Aerospace Engineering, and a Certificate in Remote Sensing in 2000. After graduating, he went to work at JSC in the Advanced Projects Office. There, he served as a lead systems engineer on several exploration studies, including those investigating the placement of an inflatable habitat at the Earth–Moon libration point, a Mars Sample Return mission, and several efforts to detail a preliminary design for a successor to the Space Shuttle. In 2003, he took a leave of absence from NASA to pursue a doctorate at UCLA in Geology/Astrobiology. As a fellow at the Center for the Study of Evolution and the Origin of Life, he examined some of the most ancient fossils ever found. For his dissertation, he established a new technique for imaging microfossils three-dimensionally within their rock matrix using confocal laser scanning imagery and Raman spectroscopy.

After graduating with his doctorate in 2007, Abhishek returned to JSC to work on the Constellation Program. There, he again served as a systems engineer, helping to integrate the various vehicles in the Constellation architecture. In 2010, he served as the on-orbit phase lead for the program. In addition to his work on the Constellation program, he served as the co-chair of the JPL-led Mars Exploration Program Analysis Group's (MEPAG) human exploration subcommittee, and one of the primary editors of NASA's Design Reference Architecture for the human exploration of Mars. Upon cancellation of the Constellation program in 2010, he moved on to the private sector and currently has a leadership role working to develop and test commercial capabilities as part of NASA's Commercial Orbital Transportation Services contract.

In his spare time, Abhishek is an avid outdoorsman and adventure-seeker. He has climbed Mt Fuji, Whitney, and Kilimanjaro and has his sights set on a few more of the world's tallest peaks. He is also a private pilot and scuba-certified. In 2002, he completed high-altitude chamber training and flew 30 microgravity parabolas in a KC-135.

Cosan Unuvar

Cosan holds a B.S. in Metallurgical Engineering and Materials Science, with honor, from Middle East Technical University (METU) in Ankara, Turkey. He came to the US with a student exchange (study abroad) program and started his graduate studies afterwards.

Cosan completed his Ph.D. in Materials Science & Engineering at the University of California, Davis (UCD) in 2006. His dissertation project was about the Effects of Gravity and Electric Current in Combustion Synthesis. Cosan led a team of professors, students, industry professionals, and government agency representatives in a joint effort in two successful reduced gravity flight campaigns (KC-135) at JSC.

He has logged eight flights with about 320 reduced-gravity parabolas, including Lunar and Martian gravities. He has also led the flight certification of research hardware, which involved a multi-disciplinary analysis and preparation of an extensive Test Equipment Data Package (TEDP). Cosan also developed a similar experimental system to be used at the NASA Glenn Research Center (GRC) Drop Tower.

Cosan continued his career at a research faculty at the Colorado School of Mines, at the Advanced Combustion Synthesis & Engineering Laboratory (ACSEL), and the Center for Space Resources (CSR). His main research project was funded by NASA Langley Research Center (LRC) to develop a contingency sterilization of a Martian Sample Return (MSR) mission, successfully using combustion synthesis as a heat source. Cosan has also worked with NASA Ames Research Center (ARC) in the development of Ultra High Temperature Ceramics (UHTC) for aerospace and atmospheric re-entry applications.

Cosan has gained significant interdisciplinary expertise thanks to his work with agencies ranging from NASA to DARPA and from the NSF to the DOE – this work has involved industrial and government/military applications such as advanced novel materials synthesis and processing, nanomaterials, ISRU, reactive material structures, lightweight armor, and biomaterials.

Cosan's interests include traveling, hiking, snowboarding, swimming, soccer, volleyball, and go-karting.

Veronica Ann Zabala-Aliberto

Veronica Ann Zabala-Aliberto ("Vaza") has lived and traveled across the US and Canada. She has had the opportunity to work on NASA's Mars Exploration Rover missions, ESA's Mars Express mission, and was the Coordinator for NASA's Lunar Reconnaissance Orbiter Camera located within the Science Operations Center at Arizona State University.

As Commander of the Family Living Analysis on Mars Expedition (FLAME) from 2005 to the present, Vaza has gained expertise in lunar and Martian analog studies, human factors adaption, and education outreach. Mission highlights have included commanding international crews to the Mars Society's MDRS, during which she accumulated 150 EVA hours.

In addition to being an avid adventurer, aviation enthusiast, and noted space activist, Vaza spends most of her time educating and inspiring the next generation of scientists, engineers, and explorers and promotes robotic and human space exploration and settlement. She has been invited to give keynotes and lectures at various venues such as the NASA Summer Mars and Robotics Camp and various scientific and leisure conventions and conferences.

She currently serves as the Phoenix, Arizona Chapter President for the National Space Society and recently won an election to serve as an At-Large Director for the National Space Society. She has been a Mars Society Phoenix Chapter member, JPL Solar System Ambassador, Planetary Society State Coordinator, Yuri's Night Coordinator for Phoenix, Arizona, and Geological Society of America member.

Zabala-Aliberto has numerous publications in the fields of robotic and human space exploration and settlement and is the recipient of several prestigious awards, including: Motorola Award (1999), Counselor's Award for Adult Space Academy (2000), NASA Space Grant (2000–05), and the Explorer Award from the National Space Society.

Pavel Zagadailov

Pavel, a physician and researcher, was born and raised in Chisinau in the Republic of Moldova. He graduated from Lyceum "N. Gogol" with a B.S. in Chemistry and Biology. There, he received several nationally recognized awards of excellence, which helped him obtain a scholarship for the State Medical and Pharmaceutical University of Moldova. Pavel developed an interest in endocrinology, embryology, and biophysics during medical school, which persuaded him to pursue a residency in obstetrics and gynecology. Aside from the intense clinical curriculum, during his residency at the National Mother and Child Health Protection Center, Pavel worked on research projects studying endocrinology aspects of normal and complicated pregnancy development and pathophysiology of the feto–placentar barrier. During the last years of his residency training, he was appointed chief resident.

After his residency, Pavel was in search of additional scientific training and research experience. He moved to the US, where he worked as a postdoctoral research associate at the University of Pittsburgh, studying inflammatory biomarker profiles and endocrinological changes in patients after brain injury. His research focused on gender specificity and he developed a number of novel concepts such as mitochondria-induced neuroinflammation in patients with closed brain trauma. During his fellowship, Pavel had an opportunity to work with a number of talented researchers from the Ob/Gyn Department at Emory University, PM&R Department and Department of Pathology at University of Pittsburgh, Safar Center for Resuscitation Research.

With over 20 years of dedication to martial arts, Pavel currently holds a second-degree black belt in Taekwondo and has participated in a number of national and international championships.

234   Astronauts for Hire Overview

Luis Zea, Newsletter Editor

Luis was raised in Guatemala City in a family in which education, hard work, and ethics were of paramount importance. Early on in his childhood, he learned the basic principles of how rockets worked through educational books. Since so much of the information was in English, it became a natural incentive for him to start learning the language. As he learned about NASA, he also discovered there were other space agencies throughout the world, and this motivated him to learn German and Portuguese as well. Needless to say, space exploration became his goal and the catalyst for other goals of personal improvement.

Luis graduated with a B.S. in Mechanical Engineering from the Universidad del Valle de Guatemala and then obtained a M.S. in Aerospace Engineering from the University of Central Florida. There, he was involved with the development of the structure for a pico-satellite and its payload: a Micro-Wave Electrothermal Thruster (plasma thruster) at Cape Canaveral. While studying for his Master's, Luis conducted research at the Florida Space Institute on gas kinetics on multi-phase flow and on CO_2 sequestration for air revitalization purposes.

He is currently pursuing a Ph.D. in Bioastronautics at the University of Colorado at Boulder while working for a space life sciences research company. There, he assists principal investigators from academia and industry to develop hardware and documentation to successfully fly their experiments on board the ISS and provides operational support while the experiments are in space.

Additionally, Luis was the Crew 65 Engineer at the MDRS, where he conducted analog EVAs and research on EVA emergency scenarios. He has worked in industry as a Mechanical and Heat Transfer Engineer. Luis has been a proud Red Cross volunteer for years and is a certified Life Guard and scuba-diver.

Appendix II
Generic Suborbital Training Schedule

WEEKLY INSTRUCTION			Course Session: *Suborbital*	Week of Instruction: 1	
	MONDAY *Insert date*	TUESDAY *Insert date*	WEDNESDAY *Insert date*	THURSDAY *Insert date*	FRIDAY *Insert date*
0800 – 0850 hrs	ADM Intro to training facility	EO 01.01 History of commercial suborbital astronaut training	PC 07.01 Swim test. Part 1: 500m in 10 minutes. Swim 25m underwater	EO 12.01 High Altitude Indoctrination Theory	ADM Personal Administration
0900 – 0950 hrs	ADM Course administration, & insurance	EO 02.01 Into to Orbital Mechanics Theory	PC 07.02 Swim Test, Part 2. Jump from 10m. Tread water for 15 minutes.	EO 12.02 High Altitude Indoctrination Theory	ADM Preflight Medical
			Break		
1010 – 1100 hrs	ADM Pre-course medical	EO 03.01 Into to the Space Environment Theory	PC 08.01 Emergency Egress/ Helo-dunker	PC 13.01 HAI: Rapid and Slow decompression Profiles	ADM Media Interviews
1110 – 1200hrs	ADM Pre-course medical	EO 04.01 Space Physiology, Theory	PC 08.02 Emergency Egress/Helo-dunker	PC 14.01 Confidence flight with pressure suit to 65,000ft	ADM Preflight Preparation

238 Generic Suborbital Training Schedule

Time					
1300 – 1350 hrs	**ADM** Equipment issue. Training manuals issue.	**EO 05.01** Unusual attitude indoctrination	**PC 09.01** Emergency Egress drills/Spacecraft mock-up	**EO 15.01** Acceleration Physiology Theory	**ADM** Inflight Ingress Activities
1400 – 1450 hrs	**ADM** Facility orientation	**EO 05.02** Unusual attitude indoctrination	**PC 09.02** Emergency Egress drills/Spacecraft mock-up	**EO 15.02** Anti-G Straining Maneuver Instruction	**ADM** Launch Countdown
			Lunch break / Break		
1500 – 1550 hrs	**ADM** Meet and greet with instructors and support staff	**EO 05.03** Unusual attitude indoctrination	**EO 10.01** Survival Training Theory	**PC 16.01** AGOR & GOR Runs to 4 and 5G	**PC 17.01** SUBORBITAL FLIGHT
1600 – 1650 hrs	**ADM** Orientation to vehicle mock-up and simulators	**EO 06.01** Emergency Egress Theory	**EO 10.02** Survival Training Theory	**PC 16.02** Flight Profile Run and Distraction Exercises	**POST-FLIGHT ADMINISTRATION** and issue of Suborbital Astronaut Wings
EVENING ASSIGNMENT	Self-study/Computer-based training	Travel to SURVIVAL SYSTEMS Self-study/Computer-based training	TRAVEL TO NASTAR Self-study/Computer-based training	TRAVEL TO LAUNCH SITE	

ADM : ADMINISTRATION
EO : ENABLING OBJECTIVE
PC : PERFORMANCE CHECK

Appendix III
Generic Orbital Training Schedule

WEEKLY INSTRUCTION			Course Session:	Orbital	Week of Instruction: 1
	MONDAY *Insert date*	**TUESDAY** *Insert date*	**WEDNESDAY** *Insert date*	**THURSDAY** *Insert date*	**FRIDAY** *Insert date*
0800 – 0850 hrs	ADM Intro to training facility	EO 01.01 History of commercial orbital astronaut training	EC 08.01 Medical Training Theory, Part 1	PC 13.01 Emergency Egress Training; Helo-dunker	PC 15.01 AFF Sky-diving Theory Part 1
0900 – 0950 hrs	ADM Course administration, & insurance	EO 02.01 Orbital Mechanics Theory Part 1	EC 08.02 Medical Training Theory, Part 2	PC 13.02 Emergency Egress Training; Helo-dunker	PC 15.02 AFF Sky-diving Theory Part 2
			Break		
1010 – 1100 hrs	ADM Pre-courses medical	EO 03.01 Space Environment Theory, Part 1	SGA Physical Fitness #1 Volleyball	EO 03.02 Space Environment Theory, Part 2	PC 15.03 Sky-dive 12,000 ft. Jumps 1, 2, 3 & 4
1110 – 1200hrs	ADM Pre-course medical	EO 04.01 Space Physiology, Theory, Part 1	SGA Physical Fitness #1 Volleyball	EO 04.02 Space Physiology Theory, Part 2	PC 15.04 Sky-dive 12,000 ft. Jumps 1, 2, 3 & 4

Generic Orbital Training Schedule

Time					
			Lunch break		
1300 – 1350hrs	ADM Equipment issue. Training manuals issue.	PC 05.01 Swim test. Part 1.500m in 10 minutes. Swim underwater 25m.	EO 10.01 Life Support Systems Theory, Part 1	EO 10.02 Life Support Systems Theory, Part 2	EO 16.01 Medical Training Practical. Part 1
1400 – 1450 hrs	ADM Facility orientation	PC 05.02 Swim Test, Part 2. Jump from 10m. Tread water for 15 minutes.	EO 11.01 Spaceflight Theory. Part 1	EO 10.03 Life Support Systems Theory, Part 3	EO 16.02 Medical Training Practical. Part 2
			Break		
1500 – 1550 hrs	ADM Meet and greet with instructors and support staff	EC 06.01 Mission Architecture, theory, Part 1.	EO 12.01 Orbital Mechanics Theory. Part 1	EO 14.01 Jungle Survival Theory	EO 17.01 Sea Survival Theory
1600 – 1650 hrs	ADM Orientation to vehicle mock-up and simulators	EC 07.01 Desert Survival Theory.	EO 06.02 Mission Architecture, Theory, Part 2	EO 11.02 Spaceflight Theory. Part 2	EO 18.01 Life Support Systems Operation Practical
EVENING ASSIGNMENT	Self-study/Computer-based training	Self-study/Computer-based training	Self-study/Computer-based training	Self-study/Computer-based training	Free

ADM : ADMINISTRATION
EO : ENABLING OBJECTIVE
SGA : SMALL GROUP ACTIVITY
PC : PERFORMANCE CHECK

Index

Adams, Mike, 55, 56, 57
Almost loss of consciousness, 48
Altman, Christopher, 39, 40
Anti-G straining maneuver, 51, 66
Armstrong, Neil, 54
Astronauts for Hire, 9, 19, 20, 21, 22, 27, 30, 35, 52, 58, 63, 67, 68, 69, 71, 109, 141
    Academic modules, 71
    Orbital training program, 81
    Training Committee, 69
Atlas Aerospace, 80

Bezos, Jeff, 85, 97, 98
Bigelow Aerospace, 7, 34, 99, 100, 127–130, 167, 174, 176, 177, 181, 182
    Advanced Medical Facility, 177
    BA-330, 100, 101, 127, 128, 129, 164, 177
    BA-2100, 129
    Bigelow Aerospace Station Complex, 170
    Genesis I, 127, 176
    Genesis II, 127, 176
    Lunar base, 181
    Sundancer, 128
Bigelow, Robert T., 127
Binnie, Brian, 11, 12
Blue Origin, 89, 97, 126, 127, 147, 148
    New Shepard, 97, 98, 126, 127
Bio-Suit, 193–195
Boeing, 133, 166
    Crew Space Transportation-100, 133, 134–136, 138
Branson, Richard, 7, 30, 94, 117, 147

Canadian Space Agency, 21, 26
Commercial Spaceflight Federation, 107
Commercial Crew Development-1, 97, 134
Commercial Crew Development-2, 89, 90, 92, 97, 127,
Commercial Orbital Transportation Services, 88, 96, 104, 131
Commercial Space Launch Act, 85, 95, 108
CubeSat, 158, 159

Durda, Dan, 38, 40, 65, 142, 155

Engle, Joe, 54
European Aviation Safety Administration, 109
European Space Agency, 21, 33, 85
Excalibur Almaz, 89, 90, 104, 105

Federal Aviation Administration, 9, 29, 30, 44, 63, 64, 107, 108, 112
    Commercial Space Transportation Advisory Committee, 109
Ferrone, Kristine, 18
Flashline Mars Arctic Research Station, 15

Garriott, Richard, 46, 78, 79
Goddard Space Flight Center, 13
Govrin, Amnon, 16
Gradual onset rate, 50, 51, 52
Gravity-induced loss of consciousness, 48, 51, 52
Greason, Jeff, 94, 95

Helium-3, 179, 184, 185

Index

Icarus Lunar Observatory Base, 190
In-situ resource utilization, 180, 181, 184, 191
International Commission on Radiological Protection, 62
International Space Station, 3, 4, 5, 72, 88, 99, 123, 127, 129, 147, 163, 175
International Symposium for Personal and Commercial Spaceflight, 99, 100, 101

Jet Propulsion Laboratory, 13
Johnson Space Center, 3

Kobrick, Ryan, 19

Launch Services Purchase Act, 85, 88
Lunar Support Scientist, 192, 193

Mackay, David, 7
Mark III Miner, 186, 187
Mars Society Desert Research Station, 15, 16
Melvill, Mike, 29, 30
Musk, Elon, 5, 85, 101, 103, 131

NanoRacks, 159, 160,
NASA, 3, 4, 5, 6, 7, 13, 14, 21, 30, 31, 32, 33, 34, 35, 40, 41, 43, 54, 55, 56, 74, 76, 85, 88, 89, 92, 96, 97, 99, 104, 107, 113, 123, 127, 129, 134, 136, 138, 141, 148, 161, 163, 166
NASTAR, 14, 22, 64, 65, 66, 67, 69
Suborbital Scientist Training Program, 16, 17, 22, 65, 67, 69
National Council on Radiation Protection, 61
Newman, Dava, 195
Next Generation Suborbital Researchers Conference, 14, 16, 17, 107, 141, 142, 143, 146, 161

Office of Commercial Space Transportation, 108
Olkin, Cathy, 38, 142
Orthostatic intolerance, 74

Palaia, Joe, 17, 18, 19
Personal Spaceflight Federation, 107
Planetary Science Institute, 155
Project Horizon, 182

Rapid onset rate, 50, 51, 52
Reimuller, Jason, 21, 22, 23, 24
Reisman, Garrett, 6, 32
Research and Education Mission, 65, 127
Romberger, Todd, 20
Rutan, Burt, 10, 12, 117

Saber Astronautics, 20
Scaled Composites, 112, 114, 117, 119
Shepard, Alan, 44, 126
Shiro, Brian, 13, 14, 16, 17, 19, 20
Sierra Nevada Corporation, 89, 123, 125
Dream Chaser, 96, 123, 125
Simon Fraser University, 26
Sounding rockets, 144–146
Black Brant, 144, 146
Southwest Research Institute, 36, 37, 38, 39, 65, 107, 141, 142, 143, 161
SpaceShipOne, 9, 11, 29, 43, 63, 92, 93, 112, 117
SpaceShipTwo, 7, 34, 44, 45, 46, 58, 59, 67, 93, 112, 115, 117
Space Adventures, 99, 174
Space Shuttle, 3, 4
Space Shuttle Mockup Facility, 3, 4
SpaceX, 5, 32, 89, 101–104, 130–132
Dragon, 89, 101, 130, 131, 132, 133
Falcon-1, 101
Falcon-9, 101, 104, 131
Spaceport America, 94, 107
Special Air Service, 26
Stand test, 156
Stefanyshyn-Piper, Heide, 176
Stern, Alan, 7, 36, 37, 38, 40, 65, 107, 142, 146, 147, 148, 155, 161
Stone, William, 188, 189
Suborbital Applications Research Group, 107

Tito, Dennis, 174

United Launch Alliance, 89
United Space Alliance, 89, 92

US Postal Service, 163

Virgin Galactic, 7, 30, 32, 33, 34, 44, 48, 93, 94, 121, 143, 147, 148, 149, 161

Walker, Joe, 54
WhiteKnightTwo, 7, 93, 117, 119
Whitesides, George, 161

X-15, 43, 54, 55, 56, 113–117

XCOR, 36, 94, 120–123, 143, 161
   Atsa Suborbital Observatory, 155, 156
   Lynx, 95, 120, 121, 143, 155, 156, 161

Zabala-Aliberto, Veronica, 16, 17

GPSR Compliance

The European Union's (EU) General Product Safety Regulation (GPSR) is a set of rules that requires consumer products to be safe and our obligations to ensure this.

If you have any concerns about our products, you can contact us on

ProductSafety@springernature.com

In case Publisher is established outside the EU, the EU authorized representative is:

Springer Nature Customer Service Center GmbH
Europaplatz 3
69115 Heidelberg, Germany

www.ingramcontent.com/pod-product-compliance
Lightning Source LLC
LaVergne TN
LVHW080311260326
834688LV00038B/1063